C[illegible]mposite Mat[illegible] [illegible]

[illegible]ic[illegible]ctions

Composite Materials for Electronic Functions

D.D.L. Chung

Composite Materials Research Laboratory
State University of New York at Buffalo
Buffalo, N.Y. 14260-4400, USA

TRANS TECH PUBLICATIONS LTD
Switzerland • Germany • UK • USA

ISBN 0-87849-836-2

Volume 12 of
Materials Science Foundations
ISSN 1422-3597

***Distributed** in the Americas by*

Trans Tech Publications Inc
PO Box 699, May Street
Enfield, New Hampshire 03748
USA

Phone: (603) 632-7377
Fax: (603) 632-5611
e-mail: ttp@ttp.net
Web: http://www.ttp.net

and worldwide by

Trans Tech Publications Ltd
Brandrain 6
CH-8707 Uetikon-Zuerich
Switzerland

Fax: +41 (1) 922 10 33
e-mail: ttp@ttp.net
Web: http://www.ttp.net

Printed in the United Kingdom
by Hobbs the Printers Ltd,
Totton, Hampshire SO40 3WX

Table of Contents

Preface

Composite materials for electronic functions are reviewed, with emphasis on non-structural composites for electronic packaging and structural composites for smart structures. These composites include those with polymer, metal, cement and carbon matrices. The electronic functions are derived from the electrical, electromagnetic and thermal properties of the composites.

1. Introduction

Composite materials are traditionally designed for the mechanical properties, due to their structural applications. However, composite materials are increasingly used in non-structural applications, such as electronic packaging. Moreover, structural composite materials that are multifunctional are increasingly needed, due to the demand of smart structures and the importance of weight saving. As a consequence, structural materials that can provide electronic functions are needed. Examples of electronic functions for structural materials are the sensing of strain, damage and temperature, the shielding of electromagnetic interference (EMI), low observability, lightning protection, grounding, cathodic protection, etc. Sensing is a key ability of a smart structure. EMI shielding is needed to protect electronics from interference from radio frequency waves, such as those emitted by cellular phones. Thus, electronic functions are valuable for both non-structural and structural composite materials. This review covers the science and engineering of composite materials for electronic functions.

2. Non-structural composite materials for electronic functions

Composite materials are traditionally designed for use as structural materials. With the rapid growth of the electronic industry, composite materials are finding more and more electronic applications. Due to the vast difference in property requirements between structural composites and electronic composites, the design criteria for these two groups of composites are different. While structural composites emphasize high strength and high modulus, electronic composites emphasize high thermal conductivity, low thermal expansion, low dielectric constant, high/low electrical conductivity, and/or EMI shielding effectiveness, depending on the particular electronic application. Low density is desirable for both aerospace structures and aerospace electronics. Structural composites emphasize processability into large parts, such as panels, whereas electronic composites emphasize processability into small parts, such as stand-alone films and coatings. Due to the small size of the parts, materials cost tends to be less of a concern for electronic composites than structural composites. For example, electronic composites can use expensive fillers, such as silver particles, which serve to provide high electrical conductivity.

Electronic packaging is the main area of application of non-structural composite materials for electronic functions. This section provides background on electronic packaging, in addition to a review of the composite materials used for electronic packaging.

2.1 Background on electronic packaging

Electronic packaging refers to (1) the packaging of the integrated circuit chips (dies); (2) the interconnections (both on and off the chips) for signal transmission, power and ground; (3) the encapsulations for protecting the chips and interconnections from moisture, chlorides, and other species in the environment; (4) the heat sinks or other cooling devices needed to remove heat from the chips; (5) the power supply; and (6) the housing for electromagnetic interference (EMI) shielding. Its conventional hierarchy has the following levels [1, 2]:

Level 0: bare chip as removed from the finished wafer.

Level 1: bare chip mounted on a chip carrier (or substrate) and encapsulated as a packaged chip (category I of level 1); bare or packaged chip(s) mounted on a module, called multichip module (MCM), together with discrete components (category II of level I).

Level 2: printed circuit board (or printed wiring board, or card) with packaged chips, modules, and other components

Level 3: backplane (or mother board) into which printed circuit boards (or cards) are inserted.

Level 4: electronic module formed by integration of backplane and power supply with an outer housing.

Level 5: system formed by integration of electronic module.

During the last 20 years, most of the attention of the electronic industry was directed to level 0, which constitutes the heart of the electronics. This effort, which was centered on semiconductors, resulted in a rapid increase in the packing density of devices on a chip, as indicated by the rapid miniaturization of electronics in the last 20 years. However, this miniaturization is accompanied by large increases in the amount of interconnections associated with each chip and in the amount of heat generated by each chip. Therefore, the key to further miniaturization currently lies on levels 1, 2 and 3. The immediate goal is to package the chips and the associated interconnections in a compact way that allows for sufficient heat removal, that can withstand the thermal cycling associated with the turning on and turning off of the electronics, that protects the electronics from environmental attack, and that allows the electronics to operate at high speeds. As the power of electronics increases, the heat dissipation problem becomes even more difficult. As the speed of electronics increases, the signal delay caused by the capacitive effect of dielectric packaging materials becomes more intolerable. It is anticipated that levels 4 and 5 will soon start to dominate the picture.

The solution of the electronic packaging problem involves the devising of packaging schemes and the use of advanced materials. Both aspects of the work are important and must take place coherently. Materials are intimately tied to processing, which is directly affected by the packaging scheme. Certain packaging schemes may not be possible unless advanced materials are used. For example, a packaging scheme may require so much heat dissipation that an advanced thermal conductor must be used. Although materials play an important role in electronic packaging, most of the work on electronic packaging is concerned with packaging schemes rather than materials. This article is focused on materials.

The actual applications of materials in electronic packaging include interconnections, printed circuit boards, substrates, encapsulations, interlayer dielectrics, die attach, electrical contacts, connectors, thermal interface materials, heat sinks, solders, brazes, lids, housings, and so on. In general, the integrated circuit chips (dies) are attached to a substrate or a printed circuit board on which the interconnection lines have been written (usually by screen printing) on each layer of the multilayer substrate or board. In order to increase the interconnection density, another multilayer involving thinner layers of conductors and interlayer dielectrics may be applied to the substrate before attachment of the chip. By means of soldered joints, wires connect between electrical contact pads on the chip and electrical contact pads on the substrate or board. The chip may be encapsulated with a dielectric for protection. It may also be covered by a thermally conducting (metal) lid. The substrate (or board) is attached to a heat sink. A thermal interface material may be placed between the substrate (or board) and the heat sink to enhance the quality of the thermal contact. The whole assembly may be placed in a thermally conducting (metal) housing.

A printed circuit board is a sheet for the attachment of chips, whether mounted on substrates, chip carriers, or otherwise, and for the drawing of interconnections. It is a polymer-matrix composite that is electrically insulating and has our conductor lines (interconnections) on one or both sides. Multilayer boards have lines on each inside layer so that interconnections on different layers may be connected by short conductor columns called electrical vias. Printed circuit boards (or cards) for the mounting of pin-inserting-type packages need to have lead insertion holes punched through the circuit board. Printed circuit boards for the mounting of surface-mounting-type packages need no holes. Surface-mounting-type packages, whether with leads, leaded chip carriers, or without leads, leadless chip carriers (LLCCs), can be mounted on both sides of a circuit board (i.e., a card), whereas pin-inserting-type packages can only be mounted on one side of a circuit board. In surface mounting technology (SMT), the surfaces of conductor patterns are connected together electrically without employing holes. Solder is typically used to make electrical connections between a surface-mounting-type package (whether leaded or leadless) and a circuit board. A lead insertion hole for pin-inserting-type packages is a plated-through hole, a hole on whose wall a metal is deposited to form a conducting penetrating connection. After pin insertion, the space between the wall and the pin is filled by solder to form a solder joint. Another type of plated-through hole is a via hole, which serves to connect different conductor layers together without the insertion of a lead.

A substrate, also called a chip carrier, is a sheet on which one or more chips are attached and interconnections are drawn. In the case of a multilayer substrate, interconnections are also drawn on each layer inside the substrate, such that interconnections in different layers are connected, if desired, by electrical vias. A substrate is usually an electrical insulator. Substrate materials include ceramics (e.g., Al_2O_3, AlN, mullite, glass ceramics), polymers (e.g., polyimide), semiconductors (e.g., silicon), and metals (e.g., aluminum). The most common substrate material is Al_2O_3. As the sintering of Al_2O_3 requires temperatures greater than 1000°C, the metal interconnections need to be refractory, such as tungsten or molybdenum. The disadvantage of tungsten or molybdenum lies in the higher electrical resistivity compared to copper. In order to make use of more conductive metals (e.g., Cu, Au, Ag-Pd) as the interconnections, ceramics that sinter at temperatures below 1000°C ("low temperature") can be used in place of Al_2O_3. The competition between ceramics and polymers for substrates is increasingly keen. Ceramics and polymers are both electrically insulating; ceramics are advantageous in that they tend to have a higher thermal conductivity than polymers; polymers are advantageous in that they tend to have a lower dielectric constant than ceramics. A high thermal conductivity is attractive for heat dissipation; a low dielectric constant is attractive for a smaller capacitive effect, hence a smaller signal delay. Metals are attractive for their very high thermal conductivity compared to ceramics and polymers.

An interconnection is a conductor line for signal transmission, power, or ground. It is usually in the form of a thick film of thickness > 1 μm. It can be on a chip, a substrate or a printed circuit board. The thick film is made by either screen printing or plating. Thick film conductor pastes containing silver particles (conductor) and glass frit (binder which functions by viscous flow of glass upon heating) are widely used to from thick film conductor lines (interconnections) on substrates by screen printing and subsequent firing. These films suffer from the reduction of the electrical conductivity by the presence of the glass and the porosity in the film after firing. The choice of a metal in a thick film paste depends on the need for withstanding air oxidation in the heating encountered in subsequent processing, which can be the firing of the green thick film together with the green ceramic substrate (a process known as cofiring). It is during cofiring that bonding and sintering take place. Copper is an excellent conductor, but it oxidizes readily when heated in air. The choice of metal also depends on the temperature encountered in subsequent processing. Refractory metals, such as tungsten and molybdenum, are suitable for interconnections heated to high temperatures (> 1000°C), for example during Al_2O_3 substrate processing.

A z-axis anisotropic electrical conductor film is a film which is electrically conducting only in the z-axis, i.e., in the direction perpendicular to the plane of the film. As one z-axis film can replace a whole array of solder joints, z-axis films are valuable for solder replacement, processing cost reduction and repairability improvement in surface mount technology.

An interlayer dielectric is a dielectric film separating the interconnection layers, such that the two kinds of layers alternate and form a thin film multilayer. The dielectric is a polymer, usually spun on or sprayed; or a ceramic, usually applied by chemical vapor deposition (CVD). The most common multilayer involves polyimide as the dielectric and copper interconnections, plated, sputtered, or electron-beam deposited.

A die attach is a material for joining a die (a chip) to a substrate. It can be a metal alloy (a solder paste), a polymeric adhesive (a thermoset or a thermoplast), or a glass. Die attach materials are usually applied by screen printing. A solder is attractive in its high thermal conductivity, which enhances heat dissipation. However, its application requires the use of heat and a flux. The flux subsequently needs to be removed chemically. The defluxing process adds costs and is undesirable to the environment (the ozone layer) due to the chlorinated chemicals used. A polymer or glass has poor thermal conductity, but this problem can be alleviated by the use of a thermally conductive filler, such as silver particles. A thermoplast provides a reworkable joint, whereas a thermoset does not. Furthermore, a thermoplast is more ductile than a thermoset. Moreover, a solder suffers from its tendency to experience thermal fatigue due to the thermal expansion mismatch between the chip and the substrate and the resulting work hardening and cracking of the solder. In addition, the footprint left by a solder tends to be larger than the footprint left by a polymer, due to the ease of the molten solder to flow.

An encapsulation is an electrically insulating conformal coating on a chip for protection against moisture and mobile ions. An encapsulation can be a polymer (e.g., epoxy, polyimide, polyimide siloxane, silicone gel, Parylene, and benzocyclobutene), which can be filled with SiO_2, BN, A1N, or other electrically insulating ceramic particles for decreasing the thermal expansion and increasing the thermal conductivity [3-5]. The decrease of the thermal expansion is needed because a neat polymer typically has a much higher coefficient thermal expansion than a semiconductor chip. An encapsulation can also be a ceramic (e.g., SiO_2, Si_3Ni_4, silicon oxynitride). In the process of electronic packaging, encapsulation is a step performed after both die bonding and wire bonding, and before the packaging using a molding material. The molding material is typically a polymer, such as epoxy. However, it can also be a ceramic, such as Si_3N_4, cordierite (magnesium silicate), SiO_2, and so on. A ceramic is advantageous (compared to a polymer) in its low coefficient of thermal expansion and higher thermal conductivity, but it is much less convenient to apply than a polymer.

A lid is a cover for a chip for physical protection. The chip is typically mounted in a well in a ceramic substrate and the lid covers the well. A lid is preferably a metal because of the need to dissipate heat. It is typically joined to the ceramic substrate by soldering, using a solder preform (e.g., Au-Sn) shaped like a gasket. Due to the low coefficient of thermal expansion (e.g., Kovar, 54Fe-29Ni-17Co) is used for the lid. For the same reason, Kovar is often used for the can (housing or enclosure) in which a substrate is mounted. Although Kovar has a low coefficient of thermal expansion (5.3×10^{-6} °C^{-1} at 20-200°C), it suffers from a low thermal conductivity of 17 W m^{-1} K^{-1}.

A heat sink is a thermal conductor that serves to conduct (mainly) and radiate heat away from the circuitry. It is typically bonded to a printed circuit board. The thermal resistance of the bond and that of the heat sink itself govern the effectiveness of the heat dissipation. A heat sink that matches the coefficient of thermal expansion of the circuit board for resistance to thermal cycling.

Insufficiently fast dissipation of heat is the most critical problem that limits the reliability and performance of microelectronics [6]. The problem becomes more severe as electronics are miniaturized, so the further miniaturization of electronics is hindered by this problem. The problem also accentuates as the power (voltage and current) increases, so the power is restricted by the heat dissipation problem. Excessive heating resulted from insufficient heat dissipation causes thermal stress in the electronic package. The stress may cause warpage of the semiconductor chip (die). The problem is compounded by thermal fatigue, which results from cyclic heating and thermal expansion mismatches. Since there are many solder joints in an electronic package, solder joint failure is a big reliability problem. For these various reasons, thermal management has become a key issue within the field of electronic packaging. Thermal management refers to the use of materials (heat sinks, thermal interface materials, etc.), devices (fans, heat pipes, etc.) and packaging schemes to attain efficient dissipation of heat. The use of materials is the part that is addressed in this article.

The use of materials with high thermal conductivity and low thermal expansion for heat sinks, lids, housings, substrates and die attach is an important avenue for alleviating the heat dissipation problem. For this purpose, metal-matrix composites (such as silicon carbide particle aluminum-matrix composites) and polymer-matrix composites (such as silver particle filled epoxy) have been developed. However, another avenue, which has received much less attention, is the improvement of the thermal contact between the various components in an electronic package (e.g., between substrate and heat sink). The higher thermal conductivity of the individual components cannot effectively help heat dissipation unless thermal contacts between the components is good. Without good thermal contacts, the use of expensive thermal conducting materials for the components is a waste.

The improvement of a thermal contact involves the use of a thermal interface material, such as a thermal fluid, a thermal grease (paste) or a resilient thermal conductor. A thermal fluid or grease is spread on the mating surfaces. A resilient thermal conductor is sandwiched by the mating surfaces and held in place by pressure. Thermal fluids are most commonly mineral oil. Thermal greases (pastes) are most commonly conducting particle (usually metal or metal oxide) filled silicone. Resilient thermal conductors are most commonly conducting particle filled elastomers. Out of these three types of thermal interface materials, thermal greases (based on polymers, particularly silicone) are by far most commonly used. Resilient thermal conductors are not as well developed as thermal fluids or greases.

As the materials to be interfaced are good thermal conductors (such as copper), the effectiveness of a thermal interface material is enhanced by high thermal conductivity and low thickness of the interface material and low thermal contact resistance between the interface material and each mating surface. As the mating surfaces are not perfectly smooth, the interface material must be able to flow or deform, so as to conform to the topology of the mating surfaces. If the interface material is a fluid, grease or paste, it should have a high fluidity (workability) so as to conform and to have a small thickness after mating. On the other hand, the thermal conductivity of the grease or paste increases with increasing filler content and this is accompanied by decrease in the workability. Without a filler, as in the case of an oil, the thermal conductivity is poor. A thermal interface material in the form of a resilient thermal conductor sheet (e.g., a collection of conducting fibers clung together without a binder, and a resilient polymer-matrix composite containing a thermally conducting filler) usually cannot be as thin or conformable as one in the form of a fluid, grease or paste, so its effectiveness requires a very high thermal conductivity within it.

Section 2 reviews composites for electronic packaging [7-9] by classifying them according to their matrix material. The following subsections cover polymer-matrix composites, metal-matrix composites, carbon-matrix composites and ceramic-matrix composites.

2.2 Polymer-matrix composites

Polymer-matrix composites with continuous or discontinuous fillers are used for electronic packaging and thermal management. Composites with continuous fillers (fibers, whether woven or not-woven) are used as substrates, heat sinks and enclosures. Composites with discontinuous fillers (particle or fibers) are used for die attach, electrically/thermally conducting adhesives, encapsulations, thermal interface materials and electrical interconnections (thick film conductors and z-axis conductors). Composites with discontinuous fillers can be in a paste form during processing, thus allowing application by printing (screen printing or jet printing) and injection molding. Composites with continuous fillers cannot undergo paste processing, but the continuous fillers provide lower thermal expansion and higher conductivity than discontinuous fillers.

Composites can have thermoplastic or thermosetting matrices. Thermoplastic matrices have the advantage that a connection can be reworked by heating for the purpose of repair, whereas thermosetting matrices do not allow reworking. On the other hand, controlled-order thermosets are attractive for their thermal stability and dielectric properties [10]. Polymers exhibiting low dielectric constant, low dissipation factor, low coefficient of thermal expansion, and compliance are preferred [11].

Composites can be electrically conducting or electrically insulating; the electrical conductivity is provided by a conductive filler. The composites can be both electrically and thermally conducting, as attained by the use of metal or graphite fillers; they can be electrically insulating but thermally conducting, as attained by the use of diamond, aluminum nitride, boron nitride or alumina fillers [12,13]. An electrically conducting composite can be isotropically conducting [14,15] or anisotropically conducting [16]. A z-axis conductor is an example of an anisotropic conductor.

2.2.1 Polymer-matrix composites with continuous fillers

Epoxy-matrix composites with continuous glass fibers and made by lamination are most commonly used for printed wiring boards, due to the electrically insulating property of glass fibers and the good adhesive behavior and established industrial usage of epoxy. Aramid (Kevlar) fibers can be used instead of glass fibers to provide lower dielectric constant [17]. Alumina (Al_2O_3) fibers can be used for increasing the thermal conductivity [18]. By selecting the fiber orientation and loading in the composite, the dielectric constant can be decreased and the thermal conductivity can be increased [19]. By impregnating the yarns or fabrics with a silica-based sol and subsequent firing, the thermal expansion can be reduced [20]. Matrices other than epoxy can be used. Examples are polyimide and cyanate ester [21].

For heat sinks and enclosures, conducting fibers are used, since the conducting fibers enhance the thermal conductivity and the ability to shield electromagnetic interference (EMI). EMI shielding is particularly important for enclosures [22]. Carbon fibers are most commonly used for these applications, due to their conductivity, low thermal expansion, and wide availability as a structural reinforcement. For high thermal conductivity, carbon fibers made from mesophase pitch [23-29] or copper plated carbon fibers are preferred [30-

32]. For EMI shielding, both uncoated carbon fibers [33, 34] and metal (e.g., nickel, copper) coated carbon fibers [35,36] have been used.

For avionic electronic enclosures, low density (lightweight) is essential for saving aircraft fuel. Aluminum is the traditional material for this application. Carbon fiber reinforced epoxy has been judged by consideration of mechanical, electrical, environmental, manufacturing/producibility, and design-to-cost criteria to be more attractive than aluminum, glass fiber reinforced epoxy, glass fiber reinforced epoxy with aluminum interlayer, beryllium, aluminum-beryllium and SiC particle reinforced aluminum [37]. A related application is thermal management of satellites, for which the thermal management materials need to be integrated from the satellite structure down to the electronic device packaging [38]. Continuous carbon fibers are suitable for this application due to their high thermal conductivity, low density, high strength and high modulus.

2.2.2 Polymer-matrix composites with discontinuous fillers

Polymer-matrix composites with discontinuous fillers (particles or short fibers) are widely used in electronics [39, 40], in spite of their poor mechanical properties compared to composites with continuous fibers. This is because materials in electronics do not need to be mechanically strong and discontinuous fillers enable processing through the paste form, which is particularly suitable for making films, whether standalone films or films on a substrate.

Screen printing is a common method for patterning a film on a substrate. In the case of an electrically conducting paste, the pattern is commonly an array of electrical interconnections and electrical contact pads on the substrate. As screen printing involves the paste going through a screen, screen printable pastes usually contain particles and no fiber and the particles must be sufficiently small, typically less than 10 μm in size. The larger the particles, the poorer is the patternability, i.e., the edge of a printed line is not sufficiently well-defined. In applications not requiring patternability, such as thermal interface materials, short fibers are advantageous in that the connectivity of the short fibers are superior to that of particles at the same volume fraction. For a conducting composite, better connectivity of the filler units means higher conductivity for the composite. Instead of using short fibers, one may use elongated particles or flakes for the sake of the connectivity. In general, the higher the aspect ratio, the better is the connectivity for the same volume fraction. The use of elongated particles or flakes can provide an aspect ratio larger than 1, while retaining patternability. Thus, it is an attractive compromise.

In case of a conducting composite, the greater the volume fraction of the conducting filler, the higher is the conductivity of the composite, since the polymer matrix is usually insulating. However, the greater the filler volume fraction, the higher is the viscosity of the paste and the poorer is the processability of the paste. To attain a high filler volume fraction while maintaining processability, a polymer of low viscosity is preferred and good wettability of the filler by the matrix, as provided by filler surface treatments and/or the use of surfactants, is desirable.

The matrix used in making a polymer-matrix composite can be in the form of a liquid (e.g., a thermosetting resin) or a solid (e.g., a thermoplastic powder) during the mixing of the matrix and the filler. In the case of the matrix in the form of a powder, the distribution of the filler units in the resulting composite depends on the size of the matrix power particles, as the filler units line the interface between adjacent matrix particles and the filler volume fraction needed for percolation (i.e., the filler units touching one another to form a continuous path) [41-50] decreases with increasing matrix particle size. The reaching of percolation is accompanied by a large increase in the conductivity. However, a large matrix particle size is detrimental to the processability. Therefore, a compromise is needed.

In the case of the matrix in the form of a thermoplastic powder, the percolation attained after mixing the matrix powder and the filler may be degraded or destroyed after subsequent composite fabrication involving flow of the thermoplastic under heat and pressure. Hence, in this case, a thermoplastic that flows less is preferred for attaining high conductivity in the resulting composite [51].

A less common way to attain percolation in a given direction is to apply an electric or magnetic field so as to align the filler units along the direction. For this technique to be possible, the filler units (whether in the bulk or on the surface) must be polarizable electrically or magnetically. Such alignment is one of the techniques used to produce z-axis conductors.

In percolation, the filler units touch one another to form continuous paths, but there is considerable contact resistance at the interface between the touching filler units. To decrease this contact resistance, thereby increasing the conductivity of the composite, one can increase the size of the filler unites, so that the amount of interface area is decreased, provided that percolation is maintained. A less common but even more effective way is to bond the filler units together at their junction by using a solid (like solder) that melts and wets the surface of the filler during the composite fabrication. The low melting point solid can be in the form of particles added to the composite mix, or in the form of a coating on the filler units. In this way, a three-dimensionally interconnected conducting network is formed after composite fabrication [52].

An intimate interface between the filler and the matrix is important to the conductivity of a composite, even though the filler is conducting and the matrix may be perfectly insulating. This is because conduction may involve a path from a filler unit to an adjacent one through a thin film of the matrix by means of tunneling. In the case of the matrix being slightly conducting (but not as conducting as the filler), the conduction path involves both the filler and the matrix and the filler-matrix interface is even more important. This interface may be improved by filler surface treatments (by the use of chemicals, heat, plasma, etc.) prior to incorporating the filler in the composite, or by the use of a surfactant [53].

The difference in thermal expansion coefficient between filler and matrix and the fact that composite fabrication occurs at an elevated temperature cause thermal stress during cooling of the fabricated composite. The thermal expansion coefficient of a polymer is usually relatively high, so the filler units are usually under compression after cooling. The compression helps to tighten the filler-matrix interface, though the compressive stress in the filler and the tensile stress in the matrix may degrade the performance and durability of the composite.

In case of the matrix being conducting, but not as conducting as the filler, as for conducting polymer matrices [54, 55], percolation is not essential for the composite to be conducting, though percolation would greatly enhance the conductivity. Below the percolation threshold (i.e., the filler volume fraction above which percolation occurs), the conductivity of the composite is enhanced by a uniform distribution of the filler units, since the chance of having a conduction path that involves more filler and less matrix increases as the filler distribution becomes more uniform. Uniformity is never perfect; it is described by the degree of dispersion of the filler. The degree of dispersion can be enhanced by rigorous agitation during mixing of the filler and matrix or by the use of a dispersant (commonly a surfactant). In the case of the matrix in the form of particles that are coarser than the filler units, the addition of fine particles to the mix also helps the dispersion of the filler [56].

Because the thermal expansion coefficient of a polymer is relatively high, the polymer matrix expands more than the filler during heating of a polymer-matrix composite. This results in the proximity between adjacent filler units changing with temperature, thus decreasing the conductivity of the composite [57]. This phenomenon is detrimental to the thermal stability of the composites. On the other hand, this phenomenon is advantageously used in temperature-sensitive switches in which the resistance increases abruptly at a certain elevated temperature [58-60].

Voltage-sensitive switching in which the conductivity abruptly increases at a certain high voltage is useful for electrostatic discharge protection. The switching is made possible by a nonlinear dependence of the electrical conductivity on the voltage. The nonlinearity can be attained in a polymer-matrix composite containing conductive and semiconductive particles [61].

Corrosion and surface oxidation of the filler are the most common causes of degradation which decreases the conductivity of the composite. Thus, oxidation resistant fillers are essential. Silver and gold are oxidation resistant, but copper is not. Due to the high cost of silver and gold, the coating of copper, nickel, hollow microspheres or other lower cost fillers by gold or silver is common for improving the oxidation resistance and conductivity [62]. By far, the most common filler is silver particles [63-65].

A z-axis anisotropic electrical conductor film is a film which is electrically conducting in the direction perpendicular to the film, but is insulating in all other directions. This film is technologically valuable for use as an interconnection material in electronic packaging (chip-to-package, package-to-board and board-to-board), as it electrically connects the electrical contact pads touching one side of the film with the corresponding contact pads touching the directly opposite side of the film. Even though the film is in one piece, it contains numerous z-axis conducting paths (not necessarily in a regular array, can be randomly distributed), so that it can provide numerous interconnections. If each contact pad is large enough to span a few z-axis conducting paths, no alignment is needed between the contact pad array and the z-axis film, whether the conducting paths are ordered or random in their distribution [66-80]. Under this situation, in order to attain a high density of interconnections, the cross section of each z-axis conducting path must be small. However, if each contact pad is only large enough to span one z-axis conducting path, alignment is needed between the contact pad array and the z-axis film, and this means that the conducting paths in the z-axis film must be ordered in the same way as the contact pad array [81]. An example of an application of a z-axis conductor film is in the interconnections between the leads from (or contact pads on) a surface mount electronic device and the contact pads on the substrate beneath the device. In this application, one piece of z-axis film can replace a whole array of solder joints, so processing cost can be much reduced. Furthermore, the problem of thermal fatigue of the solder joints can be avoided by this replacement. Another example is in the vertical interconnections in three-dimensional electronic packaging.

A z-axis film is a polymer-matrix composite containing conducting units which form the z-axis conducting paths. The conducting units are usually particles, such as metal particles and metal coated polymer particles. The particles can be clustered so that each cluster corresponds to one conducting path [66-68, 72], metal columns [73], metal particle columns (e.g., gold plated nickel) [71, 72] and individual metal coated polymer particles [70] had been used to provide z-axis conducting paths. Particle columns were formed by magnetic alignment of the particles. Using particle columns, Ref. 67 attained a conducting path width of 400 μm and a pitch (center-to-center distance between adjacent conducting paths) of 290 μm. Also using particle columns, Ref. 71 and 72 attained a conducting path width of ~ 10 μm and a pitch of ~ 100 μm. In general, a large conducting path width is desirable for decreasing the resistance per path, while a small pitch is desirable for high density interconnection. In contrast to the use of metal wires, metal columns or metal particle columns, Ref.

82 used one metal particle per conducting path (i.e., per connection). The concept of one particle per path had been demonstrated by Ref. 70 by using metal coated polymer particles. However, due to the high resistivity of the metal coating compared to the bulk metal, the z-axis resistivity of the film was high (0.5 Ω.cm for a conducting path). By using metal particle in place of metal coated polymer particles, Ref. 82 decreased the z-axis resistivity of a conducting path to 10^{-6} Ω.cm. Furthermore, Ref. 82 did not rely on a polymer (whether the matrix or the particles) for providing resilience, as the resilience is provided by the metal particles, which protrude from both sides of the standalone film. As a result, the problem of stress relaxation of the polymer is eliminated. In addition, the protrusion of the metal particles eliminates the problem of open circuiting the connection upon heating due to the higher thermal expansion of the polymer compared to the conductor [79].

Most work on z-axis adhesive films [77, 78] used an adhesive with randomly dispersed conductive particles (8-12 μm diameter) suspended in it. The particles were phenolic spheres that had been coated with nickel. After bonding under heat (180-190°C) and pressure (1.9 MPa), a particle became oval in shape (4 μm thick). There was one particle per conducting path. The main drawback of this technology is the requirement of heat and pressure for curing the adhesive. Heat and pressure are not desirable in practical use of the z-axis adhesive. Ref. 82 removed the need for heat and pressure through the choice of the polymer.

A different kind of z-axis adhesive film [81] used screening or stenciling to obtain a regular two-dimensional array of silver filled epoxy conductive dots, but this technology suffers from the large pitch (1500 μm) of the dots and the consequent need for alignment between z-axis film and contact pad array. In the work of Ref. 82, the pitch of the conducting paths in the z-axis adhesive film is as low as 64 μm.

Capacitors require materials with a high dielectric constant. Such materials in the form of thick films allow capacitors to be integrated with the electronic packaging, thereby allowing further miniaturization, in addition to performance and reliability improvements [83]. These thick film pastes involve ceramic particles with a high dielectric constant, such as barium titanate ($BaTiO_3$), and a polymer (e.g., epoxy) [84-86].

Inductors are needed for transformers, DC/DC converters and other power supply applications. They require magnetic materials. Such materials in the form of thick films allow inductors and transformers to be integrated with the electronic packaging, thereby allowing further miniaturization. These thick film pastes involve magnetic particles (e.g., ferrite) and a polymer [87-90].

The need for electromagnetic interference (EMI) shielding is increasingly rapidly due to the interference of radio frequency radiation (such as that from a cellular phone) with digital electronics, and the increasing dependence of society on digital electronics. The associated electronic pollution is an interference problem.

EMI shielding is achieved by using electrical conductors, such as metals and conductive filled polymers [62, 91-99]. EMI shielding gaskets [100-111] are resilient conductors. They are needed to electromagnetically seal an enclosure. The resilient conductors are most commonly elastomers (e.g., rubber) that are filled with a conductive filler [112], or elastomers that are coated with a metalized layer. Metallized elastomers suffer from poor durability due to the tendency of the metal layer to debond from the elastomer. Conductive filled elastomers do not have this problem, but they require the use of a highly conductive filler, such as silver particles, in order to attain a high shielding effectiveness while maintaining resilience. The highly conductive filler tends to be expensive, making the composite expensive. The use of a less conducting filler results in the need for a large volume fraction of the filler in order to attain a high shielding effectiveness; the consequence is diminished resilience or even loss of resilience. Moreover, these composites suffer from degradation of the shielding effectiveness in the presence of moisture or solvents. In addition, the polymer matrix in the composites limits the temperature resistance, and the thermal expansion mismatch between filler and matrix limits the thermal cycling resistance.

Due to the skin effect (i.e., electromagnetic radiation at a high frequency interacting with only the near surface region of an electrical conductor), a filler for a polymer-matrix composite for EMI shielding needs to be not only electrically conducting, but also small in unit size. Although connectivity between the filler units is not required for shielding, it helps. Therefore, a filler in the form of a metal fiber of very small diameter is desirable. For this purpose, nickel filaments of diameter 0.4 μm and length > 100 μm, with a carbon core of diameter 0.1 μm, were developed [113]. Their exceptionally small diameter compared to those of existing metal fibers made them outstanding for use as a filler in a polymer for EMI shielding. A shielding effectiveness of 87 dB at 1 GHz was attained in a polyethersulfone-matrix composite with only 7 vol.% nickel filaments [113-115]. The low volume fraction allows resilience in a silicone-matrix composite for EMI gaskets [116].

Electrically insulating but thermally conducting composites are attractive for heat dissipation. Diamond is the best material for this purpose, but it is expensive. Polymer-matrix composites containing electrically insulating but thermally conducting fillers are thus attractive. Examples of such fillers are BN and AIN particles [12, 13, 117-119].

2.3 Metal-matrix composites

Metal-matrix composites [120, 121] are mostly used in electronic packaging due to their combination of high thermal conductivity (due to the metal matrix) and low coefficient of thermal expansion (due to the filler). The main applications are heat sinks, substrates, lids, housing (enclosures) and baseplates for power modules [122-131]. As the filler usually has lower thermal expansion and lower thermal conductivity than the metal matrix, the higher the filler volume fraction in the composite, the lower the coefficient of thermal expansion and the lower is the thermal conductivity.

Composites with discontinuous fillers (commonly particles) are attractive for their processability into various shapes. However, layered composites in the form of a matrix-filler-matrix sandwich are useful for planar components. Discontinuous fillers are most commonly ceramic particles (e.g., SiC). The filler sheets are most commonly low thermal expansion metal alloy sheets (e.g., Invar or 64Fe-36Ni, and Kovar or 54Fe-29Ni-17Co). The coefficient of thermal expansion (CTE) is 5 x 10^{-6}/°C, 1.6 x 10^{-6}/°C and 5.1 x 10^{-6}/°C for SiC, Invar and Kovar respectively. As Invar and Kovar are poor in thermal conductivity, Cu-Fe-Ni alloys that exhibit higher thermal conductivity are emerging [132]. Aluminum and copper are the dominant metal matrices due to their high conductivity (especially that of copper). In the case of aluminum, it is also due to the low density, which is important for avionic electronics and applications (e.g., laptop computers) which require low weight [133, 134]. Less common metal matrices include magnesium, beryllium and solder.

2.3.1 Aluminum-matrix composites

Aluminum is the dominant matrix for metal-matrix composites for both structural and electronic applications. This is because of the low cost of aluminum and the low melting point of aluminum (660°C) facilitating composite fabrication by methods that involve the melting of the metal.

Liquid-phase methods for the fabrication of metal-matrix composites include liquid metal infiltration, which usually involves using pressure (from a piston or compressed gas) to push the molten metal into the pores of a porous preform comprising the filler (commonly particles that are not sintered) and a small amount of a binder [135-137]. Pressureless infiltration is less common but is possible [138, 139]. The binder prevents the filler particles from moving during the infiltration, and also provides sufficient compressive strength to the preform, so that the preform will not be deformed during the infiltration. This method thus provides near-net-shape fabrication, i.e., the shape and size of the composite product are the same of those of the preform. Since machining of the composite is far more difficult than that of the preform, near-net-shape fabrication is desirable.

In addition to near-net-shape fabrication capability, liquid metal fabrication is advantageous in being able to provide composites with high filler volume fractions (up to 70%). A high filler volume fraction is necessary in order to attain a low enough coefficient of thermal expansion (CTE) (< 10 x 10^{-6} /°C) in the composite, even if the filler is a low CTE ceramic (e.g., SiC), since the aluminum matrix has a relatively high CTE [140, 141]. However, to attain a high volume fraction using liquid metal infiltration, the binder used must be in a small amount (so as not to clog the pores in the preform) and still be effective. Hence, the binder technology [142-144] is critical.

The ductility of a composite decreases as the filler volume fraction increases, so that a composite with a low enough CTE is quite brittle. Although the brittleness is not acceptable for structural applications, it is acceptable for electronic applications.

Another liquid-phase technique is stir casting [120], which involves stirring the filler in the molten metal and then casting. This method suffers from the non-uniform distribution of the filler in the composite due to the difference in density between filler and molten metal and the consequent tendency for the filler to either float or sink in the molten metal prior to solidification. Stir casting also suffers from the incapability of producing composites with a high filler volume fraction.

Yet another liquid-phase technique is plasma spraying [145], which involves spraying a mixture of molten metal and filler onto a substrate. This method suffers from the relatively high porosity of the resulting composite and the consequent need for densification by hot isostatic pressing or other methods, which tend to be expensive.

A solid-phase technique is powder metallurgy, which involves mixing the matrix metal powder and filler and subsequent sintering under heat and pressure [145]. This method is relatively difficult for the aluminum matrix because aluminum has a protective oxide on it and the oxide layer on the surface of each aluminum particle hinders sintering. Furthermore, this method is usually limited to low volume fractions of the filler.

The most common filler used is silicon carbide (SiC) particles, due to the low cost and low CTE of SiC [146-150]. However, SiC suffers from its reactivity with aluminum. The reaction is

$$3SiC + 4Al \rightarrow 3Si + Al_4C_3$$

It becomes more severe as the composite is heated. The aluminum carbide is a brittle reaction product which lines the filler-matrix interface of the composite, thus weakening the interface. Silicon, the other reaction product, dissolves in the aluminum matrix, lowering the melting temperature of the matrix, and causing non-uniformity in the phase distribution and mechanical property distribution [151]. Furthermore, the reaction consumes a part of the SiC filler [152].

A way to diminish this reaction is to use an Al-Si alloy matrix, since the silicon in the alloy matrix promotes the opposite reaction and is thus against this reaction. However, the Al-Si matrix is less ductile than the Al matrix, thus causing the mechanical properties of the Al-Si matrix composite to be very poor compared to those of the corresponding Al-matrix composite. Thus, the use of an Al-Si alloy matrix is not a solution to the problem.

An effective solution is to replace Si-C by aluminum nitride (AlN) particles, which do not react with aluminum, thus resulting in superior mechanical properties in the composite [150, 153]. The fact that AlN tends to have a higher thermal conductivity than SiC helps the thermal conductivity of the composite. Since the cost of the composite fabrication process dominates the cost of producing the composites, the higher material cost of AlN compared to SiC does not matter, especially for electronic packaging. Aluminum oxide (Al_2O_3) also does not react with aluminum, but it is low in thermal conductivity and tends to suffer from particle agglomeration [153].

Other than ceramics such as SiC and AlN, another filler used in aluminum-matrix composites is carbon in the form of fibers of diameter around 10 μm [154-158] and less commonly, filaments of diameter less than 1 μm [159]. Carbon also suffers from reactivity with aluminum to form aluminum carbide. However, fibers are more effective than particles for reducing the CTE of the composite. Carbon fibers can be even continuous in length. Moreover, carbon, especially when graphitized, is much more thermally conductive than ceramics. In fact, carbon fibers that are sufficiently graphitic are even more thermally conductive than the composite increases with increasing fiber volume fraction. However, these fibers are expensive. The mesophase-pitch-based carbon fiber K-1100 from Amoco Performance Products (Alpharetta, GA) exhibits longitudinal thermal conductivity 1000 W/m.K [160, 161].

Both carbon and SiC suffer from forming a galvanic couple with aluminum, which is the anode – the component in the composite that is corroded. The corrosion becomes more severe in the presence of heat and/or moisture.

The thermal conductivity of aluminum-matrix composites depends on the filler and its volume fraction, the alloy matrix heat treatment condition, as well as the filler-matrix interface [153, 162].

To increase the thermal conductivity of SiC aluminum-matrix composite, a diamond film can be deposited on the composite [163]. The thermal conductivity of single crystal diamond is 2000-2500 W/m.K, though a diamond film is not single crystalline.

2.3.2 Copper-matrix composites

Copper (398 W/m.K) is much more thermally conductive than aluminum (247 W/m.K), but its density is also much higher (8.94 g/cm^3 for copper and 2.71 g/cm^3 for aluminum). Therefore, copper is an attractive matrix for electronics that are not for avionic applications.

Because copper is heavy anyway, the filler does not have to be lightweight. Thus, low CTE but heavy metals such as tungsten [164, 165], molydenum [166-168] and Invar (64Fe-36Ni alloy) [169-171] are used as fillers. The CTE of copper is 17.0 x 10^{-6}/°C, compared to 4.5 x 10^{-6}/°C for tungsten, 4.9 x 10^{-6}/°C for molybdenum, and 1.6 x 10^{-6}/°C for Invar. These metals (except Invar) have the advantage that they are more thermally conductive than most ceramics. The thermal conductivity is 178, 142 and 10 W/m.K for tungsten, molybdenum and Invar respectively. Furthermore, they are available in particle and sheet forms, so that they are suitable for particulate as well as layered [172, 173] composites. Yet another advantage of the metal fillers is the better wettability of the molten matrix metal with metal fillers than with ceramic fillers, in case the composite is fabricated by a liquid phase method.

An advantage of copper over aluminum is its nonreactivity with carbon, so carbon is a highly suitable filler for copper. Additional advantages are that carbon is lightweight and carbon fibers are available in a continuous form. Furthermore, copper is a rather noble metal, as shown by its position in the Electromotive, so it does not suffer from the corrosion which plagues aluminum. Carbon used as a filler in copper is in the form of fibers of diameter around 10 μm [157, 174-181]. As carbon fibers that are sufficiently graphitic are even more thermally conductive than copper, the thermal conductivity of a copper-matrix composite can exceed that of copper.

Less common fillers for copper are ceramics such as silicon carbide, titanium diboride (TiB_2) and alumina [182-184].

The melting point of copper is much higher than that of aluminum, so the fabrication of copper-matrix composites is commonly done by powder metallurgy, although liquid metal infiltration is also used [157, 185, 186]. In the case of liquid metal infiltration, the metal matrix is often a copper alloy (e.g., Cu-Ag) chosen for the reduced melting temperature and good castability [186].

Powder metallurgy conventionally involves mixing the metal matrix powder and the filler, and subsequent pressing and then sintering under either heat or both heat and pressure. The problem with this method is that it is limited to low volume fractions of the filler. In order to attain high volume fractions, a less conventional method of powder metallurgy is recommended. This method involves coating the matrix metal on the filler units, followed by pressing and sintering [167, 182, 187, 188]. The mixing of matrix metal powder with the coated filler is not necessary, although it can be done to decrease the filler volume fraction in the composite. The metal coating on the filler forces the distribution of matrix metal to be uniform even when the metal volume fraction is low (i.e., when the filler volume fraction is high). On the other hand, with the conventional method, the matrix metal distribution is not uniform when the filler volume fraction is high, thus causing porosity and the presence of filler agglomerates, in each of which the filler units directly touch one another; this microstructure results in low thermal conductivity and poor mechanical properties.

Continuous carbon fiber copper-matrix composites can be made by coating the fibers with copper and then diffusion bonding (i.e., sintering) [174,

176, 180, 189-192]. This method is akin to the abovementioned less conventional method of powder metallurgy.

Less common fillers used in copper include diamond powder [186, 193], aluminosilicate fibers [194] and Ni-Ti alloy rod [195]. The Ni-Ti alloy is attractive for its negative CTE of –21 x 10^{-6}/°C.

The coating of a carbon fiber copper-matrix composite with a diamond film has been done to enhance the thermal conductivity [179].

2.3.3 Magnesium-matrix composites

Magnesium has an even lower density (1.74 g/cm^3) than aluminum (2.7 g/cm^3), but its thermal conductivity is lower than aluminum and it is an active metal (poor corrosion resistance), as shown by its position in the Electromotive Series. Moreover, its CTE is slightly higher than that of aluminum. Nevertheless, the low density makes it attractive for aerospace electronics. The alloy Mg-5.5Zn-0.5Zr (called ZK60A), which has a thermal conductivity of 117 W/m.K, has been used as the matrix to make composites with diamond powder. The composites were made by coating the diamond particles with the alloy and then consolidation of the coated powder by hot pressing [196]. However, the thermal conductivity was not increased by the diamond addition, probably due to poor consolidation or poor bonding between filler and matrix.

2.3.4 Beryllium-matrix composites

Beryllium oxide (BeO) has a high thermal conductivity of 230 W/m.K. The value for aluminum nitride (AlN) is only 50-170 W.m.K. Beryllium-matrix BeO-platelet composites with 20-60 vol.% BeO exhibit low density (2.30 g/cm^3 at 40 vol.% BeO, compared to 2.9 g/cm^3 for Al/SiC at 40 vol.% SiC), high thermal conductivity (232 W/m.K at 40 vol.% BeO, compared to 130 W/m.K for Al/SiC at 40 vol.% SiC), low CTE (7.5 x 10^{-6}/°C at 40 vol.% BeO, compared to 12.1 x 10^{-6}/°C at 40 vol.% SiC), and high modulus (317 GPa at 40 vol.% BeO, compared 134 GPa for Al/SiC at 40 vol.% SiC) [197, 198]. The applications of beryllium-matrix composites are similar to those of aluminum-matrix composites.

2.3.5 Solder-matrix composites

Solder is the dominant material used for electrical interconnections. However, their CTE (24 x 10^{-6}/°C for 60Sn-40Pb solder) is high compared to those of printed wiring boards (CTE of 16 x 10^{-6}/°C) or ceramic substrates (CTE of 6 x 10^{-6}/°C), thus causing thermal fatigue, which leads to solder joint failure. Solder joint failure accounts for 70% of failures in electronic components [199]. The addition of low CTE particles (e.g., molybdenum) to form a solder-matrix composite is a way to decrease the CTE [200]. However, the difference in density between filler and matrix cause non-uniformity in the filler distribution once the solder melts during soldering. Another way to alleviate the CTE mismatch problem is to add shape memory alloy wires (NiTi coated with Ni, which promotes the wettability by the molten solder) to the solder to form a composite that provides adaptive solder joints [201]. The concept involves using the shape memory alloy to relax the thermal stress. Its feasibility still needs to be shown.

Solder also suffers from creep. To improve the creep resistance, metal particles (e.g., Cu, Ag, Au and Ni) and intermetallic particles (e.g., Cu_3Sn and Cu_6Sn_5 which form and grow during soldering to copper) have been added to solder to form a composite [202, 203].

The need for flux is another shortcoming of soldering. Composite solders in layered rather than particulate form have been developed for soldering without flux. The lead-indium-gold multilayer composite solder is deposited directly on to the semiconductor in high vaccum. The gold outer layer serves to protect the indium layer from oxidation [204].

Although solder is most commonly used in paste and wire forms, it is also used in the form of a shaped preform. Continuous carbon fibers effectively reinforce solder preforms for the purpose of reducing the CTE [205].

Closely related to soldering is brazing. The addition of short carbon fibers to a brazing alloy helps improve the strength of brazed joints to alumina. The brazing alloy used is based on silver. [206]

2.4 Carbon-matrix composites

Carbon is an attractive matrix for composites for electronic packaging and thermal management because of its thermal conductivity (though not as high as those of metals) and low CTE (lower than those of metals). Furthermore, carbon is corrosion resistant (more corrosion resistant than metals) and lightweight (lighter than metals). Yet another advantage of the carbon matrix is its compatibility with carbon fibers, in contrast to the common reactivity between a metal matrix and its fillers. Hence, carbon fibers are the dominant filler for carbon-matrix composites. Composites with both filler and matrix being carbon are called carbon-carbon composites [207]. Their primary applications in relation to electronic packaging and thermal management are heat sinks [208, 209], thermal planes [210] and substrates [211]. There is considerable competition between carbon-carbon composites and metal-matrix composites for the same applications.

The main drawback of carbon-matrix composites is their high cost of fabrication, which typically involves making a pitch-matrix or resin-matrix composite and subsequent carbonization (by heating at 1000-1500°C in an inert atmosphere) of the pitch or resin to form a carbon-matrix composite. After carbonization, the porosity is substantial in the carbon matrix, so pitch or resin is impregnated into the composite and then carbonization is carried out again. Quite a few impregnation-carbonization cycles are needed in order to reduce the porosity to an acceptable level, thus resulting in the high cost of fabrication. Graphitization (by heating at 2000-3000°C in an inert atmosphere) may follow carbonization (typically in the last cycle) in order to increase the thermal conductivity, which increases with the degree of graphitization. However, graphitization is an expensive step. Some or all of the impregnation-carbonization cycles may be replaced by chemical vapor infiltration (CVI), in which a carbonaceous gas infiltrates the composite and decomposes to form carbon.

Carbon-carbon composites have been made by using conventional carbon fibers of diameter around 10 μm [208, 210, 212], as well as carbon filaments grown catalytically from carbonaceous gases and of diameter less than 1 μm [159]. By using graphitized carbon fibers, thermally conductivities exceeding that of copper can be reached.

To increase the thermal conductivity, carbon-carbon composites have been impregnated with copper [213, 214] and have been coated with a diamond film [215].

A competing material which is akin to (but not exactly) a carbon-matrix composite is an all-carbon material (called ThermalGraph, a tradename of BP Amoco, Alpharetta, GA) made by consolidating oriented precursor carbon fibers without a binder and subsequent carbonization and optional graphitization. The thermal conductivity ranges from 390 to 750 W/m.K in the fiber direction of the material.

Another competing material which is akin to (but not exactly) a carbon-matrix composite is pyrolytic graphite (called TPG) encased in a structural shell [216]. The graphite (highly textured with the c-axes of the grains preferentially perpendicular to the plane of the graphite), has an in-plane thermal conductivity of 1700 W/m.K (four times that of copper), but it is mechanically weak due to the tendency to shear in the plane of the graphite. The structural shell serves to strengthen by hindering shear.

Pitch-derived carbon foams with thermal conductivity up to 150 W/m.K after graphitization are attractive for their high specific thermal conductivity (thermal conductivity divided by the density) [217].

2.5 Ceramic-matrix composites

Ceramic-matrix composites with SiC, glass and cement matrices have been made for use as substrates. In addition, cement-matrix composites have been made for use in EMI shielding.

The SiC matrix is attractive due to its high CTE compared to the carbon matrix, though it is not as thermally conductive as carbon. The CTE of carbon-carbon composites is too low ($0.25 \times 10^{-6}/^{o}C$), thus resulting in reduced fatigue life in chip-on-board (COB) applications with silica chips (CTE = $2.6 \times 10^{-6}/^{o}C$). The SiC-matrix carbon fiber composite is made from a carbon-carbon composite by converting the matrix from carbon to SiC [212]. To improve the thermal conductivity of the SiC-matrix composite, coatings in the form of chemical vapor deposited AlN or Si have been used. The SiC-matrix metal (Al or Al-Si) composite, as made by a liquid-exchange process, also exhibits relatively high thermal conductivity [218].

The borosilicate glass matrix is attractive due to its low dielectric constant (4.1 at 1 MHz for B_2O_3-SiO_2-Al_2O_3-Na_2O glass), compared to 8.9 for AlN, 9.4 for alumina (90%), 42 for SiC, 6.8 for BeO, 7.1 for cubic boron nitride, 5.6 for diamond and 5.0 for glass-ceramic. However, glass has a low thermal conductivity. Hence, fillers with relatively high thermal conductivity are used with the glass matrix. An example is continuous SiC fibers, the glass-matrix composites of which are made by tape casting, followed by sintering [219]. Another example is aluminum nitride with interconnected pores (about 28 vol.%), the composites of which are obtained by glass infiltration to a depth of about 100 μm [219-221].

Polymer-impregnated cement (specifically polymer-impregnated macro-defect-free cement) exhibit attractive dielectric and mechanical properties for use as low-cost substrates [222]. Cement-matrix composites containing discontinuous carbon fibers are electrically conducting and are attractive for use in EMI shielding [223, 224].

Another form of glass-matrix composite is a thick film dielectric material (with glass as the binder) containing hollow microspheres for the purpose of reducing the dielectric constant [225].

Multilayer ceramic substrates (most commonly alumina) with metal conductor lines (e.g., tungsten) between the layers is a layered composite that is extensively used in electronic packaging [226]. It is made by sintering a multilayer consisting of ceramic green tape with metal thick film patterned on the tape.

3. Structural composite materials for electronic functions

Smart structures are important due to their relevance to hazard mitigation, structural vibration control, structural health monitoring, transportation engineering and thermal control. Research on smart structures has emphasized the incorporation of various devices in a structure for providing sensing, energy dissipation, actuation, control or other functions. Research on smart composites has emphasized the incorporation of a smart material in a matrix material for enhancing the smartness or durability. Research on smart materials has emphasized the study of materials (e.g., piezoelectric materials) used for making the devices. However, relatively little attention has been given to the development of structural materials (e.g., concrete and composites) that are inherently able to provide some of the smart functions, so that the need for embedded or attached devices is reduced or eliminated, thereby lowering cost, enhancing durability, increasing the smart volume, and minimizing mechanical property degradation (which usually occurs in the case of embedded devices).

Smart structures are structures that have the ability to sense certain stimuli and be able to response to the stimuli in an appropriate fashion, somewhat like a human being. Sensing is the most fundamental aspect of a smart structure. A structural composite which is itself a sensor is multifunctional.

Section 3 is focused on structural composites for smart structures. It addresses cement-matrix and polymer-matrix composites. The smart functions addressed include strain sensing (for structural vibration control and traffic monitoring), damage sensing (both mechanical and thermal damage in relation to structural health monitoring), temperature sensing (for thermal control, hazard mitigation and structural performance control), and vibration reduction (for structural vibration control).

3.1 Cement-matrix composites for smart structures

Cement-matrix composites include concrete (containing coarse and fine aggregates), mortar (containing fine aggregate but no coarse aggregate) and cement paste (containing no aggregate, whether coarse or fine). Other fillers, called admixtures, can be added to the mix to improve the properties of the composite. Admixtures are discontinuous, so that they can be included in the mix. They can be particles, such as silica fume (a fine particulate) and latex (a polymer in the form of a dispersion). They can be short fibers, such as polymer, steel, glass or carbon fibers. They can be liquids such as methylcellulose aqueous solution, water reducing agent, defoamer, etc. Admixtures for rendering the composite smart while maintaining or even improving the structural properties are the focus of Section 3.1.

3.1.1 Background on cement-matrix structural composites

Cement-matrix composites for smart structures include those containing short carbon fibers (for sensing strain, damage and temperature and for thermal control), short steel fibers (for sensing temperature and for thermal control) and silica fume (for vibration reduction). This section provides background on cement-matrix composites, with emphasis on carbon fiber cement-matrix composites due to its dominance among intrinsically smart cement-matrix composites.

Carbon fiber cement-matrix composites are structural materials that are gaining in importance quite rapidly due to the decrease in carbon fiber cost [227] and the increasing demand of superior structural and functional properties. These composites contain short carbon fibers, typically 5 mm in length, as the short fibers can be used as an admixture in concrete (whereas continuous fibers cannot be simply added to the concrete mix) and short fibers are less expensive than continuous fibers. However, due to the weak bond between carbon fiber and the cement matrix, continuous fibers [228-230] are much more effective than short fibers in reinforcing concrete. Surface treatment of carbon fiber (e.g., by heating [231] or by using ozone [232, 233], silane [234], SiO_2 particles [235] or hot NaOH solution [236]) is useful for improving the bond between fiber and matrix, thereby improving the properties of the composite. In the case of surface treatment by ozone or silane, the improved bond is due to the enhanced wettability by water. Admixtures such as latex [232, 237] methylcellulose [232] and silica fume [238] also help the bond.

The effect of carbon fiber addition on the properties of concrete increases with fiber volume fraction [239], unless the fiber volume fraction is so high that the air void content becomes excessively high [240]. (The air void content increases with fiber content and air voids tend to have a negative effect on many properties, such as the compressive strength.) In addition, the workability of the mix decreases with fiber content [239]. Moreover, the cost increases with fiber content. Therefore, a rather low volume fraction of fibers is desirable. A fiber content as low as 0.2 vol. % is effective [241], although fiber contents exceeding 1 vol. % are more common [242-246]. The required fiber content increases with the particle size of the aggregate, as the flexural strength decreases with increasing particle size [247].

Effective use of the carbon fibers in concrete requires dispersion of the fibers in the mix. The dispersion is enhanced by using silica fume (a fine particulate) as an admixture [240, 248-250]. A typical silica fume content is 15% by weight of cement [240]. The silica fume is typically used along with a small amount (0.4% by weight of cement) of methylcellulose for helping the dispersion of the fibers and the workability of the mix [240]. Latex (typically 15-20% by weight of cement) is much less effective than silica fume for helping the fiber dispersion, but it enhances the workability, flexural strength, flexural toughness, impact resistance, frost resistance and acid resistance [240, 251, 252]. The ease of dispersion increases with decreasing fiber length [250].

The improved structural properties rendered by carbon fiber addition pertain to the increased tensile and flexible strengths, the increased tensile ductility and flexural toughness, the enhanced impact resistance, the reduced drying shrinkage and the improved freeze-thaw durability [239-241, 243-251, 253-264]. The tensile and flexural strengths decrease with increasing specimen size, such that the size effect becomes larger as the fiber length increases [265]. The low drying shrinkage is valuable for large structures and for use in repair [266, 267] and in joining bricks in a brick structure [268, 269]. The functional properties rendered by carbon fiber addition pertain to the strain sensing ability [233, 270-284] (for smart structures), the temperature sensing ability [285-288], the damage sensing ability [270, 274, 289-291], the thermoelectric behavior [286-288], the thermal insulation ability [292-294] (to save energy for buildings), the electrical conduction ability [223, 295-303] (to facilitate cathodic protection of embedded steel and to provide electrical grounding or connection), and the radio wave reflection/absorption ability [224, 304-307] (for electromagnetic interference or EMI shielding, for lateral guidance in automatic highways, and for television image transmission).

In relation to the structural properties, carbon fibers compete with glass, polymer and steel fibers [244, 253-255, 258, 262-264, 308]. Carbon fibers (isotropic pitch based) [227, 308] are advantageous in their superior ability to increase the tensile strength of concrete, even though the tensile strength, modulus and ductility of the isotropic pitch based carbon fibers are low compared to most other fibers. Carbon fibers are also advantageous in the relative chemical inertness [309]. PAN-based carbon fibers are also used [243, 245, 248, 259], although they are more commonly used as continuous fibers than short fibers. Carbon-coated glass fibers [310, 311] and submicron diameter carbon filaments [302-304] are even less commonly used, although the former is attractive for the low cost of glass fibers and the latter is attractive for its high radio wave reflectivity (which results from the skin effect). C-shaped carbon fibers are more effective for strengthening than round carbon fibers [312], but their relatively large diameter makes them less attractive. Carbon fibers can be used in concrete together with steel fibers, as the addition of short carbon fibers to steel fiber reinforced mortar increases the fracture toughness of the interfacial zone between steel fiber and the cement matrix [313]. Carbon fibers can also be used in concrete together with steel bars [314, 315], or together with carbon fiber reinforced polymer rods [316].

In relation to most functional properties, carbon fibers are exceptional compared to the other fiber types. Carbon fibers are electrically conducting, in contrast to glass and polymer fibers, which are not conducting. Steel fibers are conducting, but their typical diameter (≥ 60 μm) is much larger than the diameter of a typical carbon fiber (15 μm). The combination of electrical conductivity and small diameter makes carbon fibers superior to the other fiber types in the area of strain sensing and electrical conduction. However, carbon fibers are inferior to steel fibers for providing thermoelectric composites, due to the high electron concentration in steel and the low hole concentration in carbon.

Although carbon fibers are thermally conducting, addition of carbon fibers to concrete lowers the thermal conductivity [292], thus allowing applications related to thermal insulation. This effect of carbon fiber addition is due to the increase in air void content. The electrical conductivity of carbon fibers is higher than that of the cement matrix by about 8 orders of magnitude, whereas the thermal conductivity of carbon fibers is higher than that of the cement matrix by only one or two orders of magnitude. As a result, the electrical conductivity is increased upon carbon fiber addition in spite of the increase in air void content, but the thermal conductivity is decreased upon fiber addition.

The use of pressure after casting [317], and extrusion [318, 319] can result in composites with superior microstructure and properties. Moreover, extrusion improves the shapability [319].

3.1.2 Cement-matrix composites for strain sensing

The electrical resistance of strain sensing concrete (without embedded or attached sensors) changes reversibly with strain, such that the gage factor (fractional change in resistance per unit strain) is up to 700 under compression or tension [233, 270-284]. The resistance (DC/AC) increases reversibly upon tension and decreases reversibly upon compression, due to fiber pull-out upon microcrack opening (< 1 μm) and the consequent increase in fiber-matrix contact resistivity. The concrete contains as low as 0.2 vol.% short carbon fibers, which are preferably those that have been surface treated. The fibers do not need to touch one another in the composite. The treatment improves the wettability with water. The presence of a large aggregate decreases the gage factor, but the strain sensing ability remains sufficient for practical use. Strain sensing concrete works even when data acquisition is wireless. The applications include structural vibration control and traffic monitoring.

Fig. 1(a) shows the fractional change in resistivity along the stress axis as well as the strain during repeated compressive loading at an increasing stress amplitude for carbon-fiber latex cement paste at 28 days of curing. Fig. 1(b) shows the corresponding variation of stress and strain during the repeated loading.

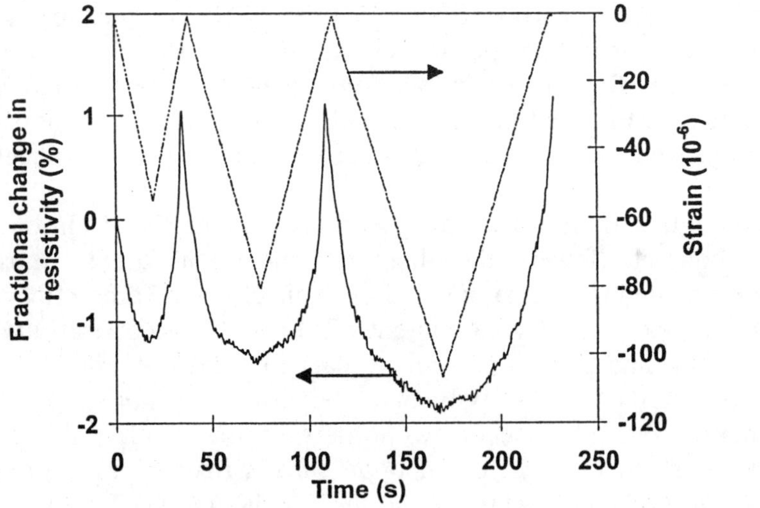

(a)

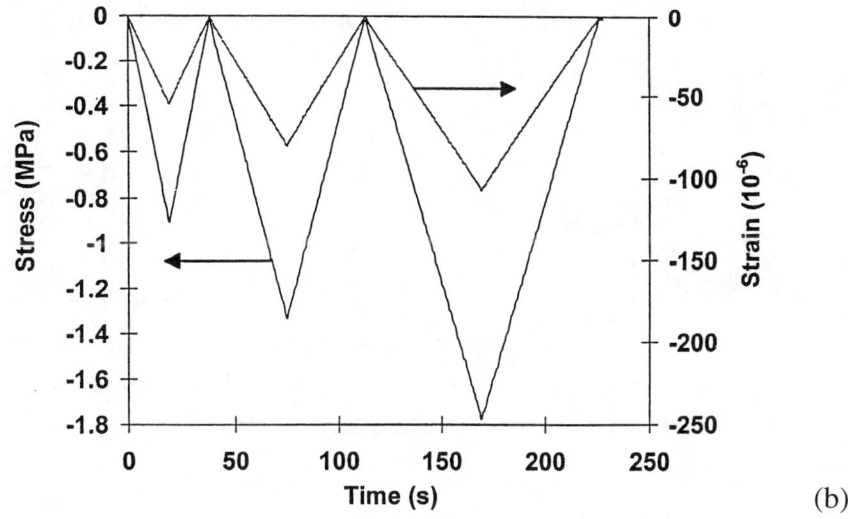

(b)

Fig. 1 *Variation of the fractional change in volume electrical resistivity with time (a), of the stress with time (b), and of the strain (negative for compressive strain) with time (a,b) during dynamic compressive loading at increasing stress amplitudes within the elastic regime for carbon-fiber latex cement paste at 28 days of curing.*

The strain varies linearly with the stress up to the highest stress amplitude (Fig. 1(b)). The strain returns to zero at the end of each cycle of loading. The resistivity decreases upon loading in every cycle (due to fiber push-in) and increases upon unloading in every cycle (due to fiber pull-out). The resistivity has a net increase after the first cycle, due to damage. Little further damage occurs in subsequent cycles, as shown by the resistivity after unloading not increasing much after the first cycle. The greater the strain amplitude, the more is the resistivity decrease during loading, although the resistivity and strain are not linearly related. The effects of Fig. 1 were similarly observed in carbon-fiber silica-fume cement paste at 28 days of curing.

Figures 2 and 3 show the fractional changes in the longitudinal and transverse resistivities respectively for carbon-fiber silica-fume cement paste at 28 days of curing during repeated unaxial tensile loading at increasing strain amplitudes. The strain essentially returns to zero at the end of each cycle, indicating elastic deformation. The longitudinal strain is positive (i.e., elongation); the transverse strain is negative (i.e., shrinkage due to the Poisson Effect). Both longitudinal and transverse resistivities increase reversibly upon

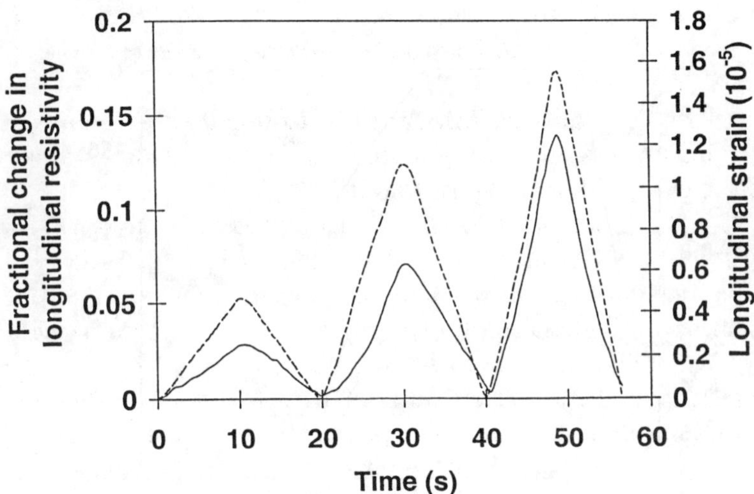

Fig. 2 *Variation of the fractional change in longitudinal electrical resistivity with time (solid curve) and of the strain with time (dashed curve) during dynamic uniaxial tensile loading at increasing stress amplitudes within the elastic regime for carbon-fiber silica-fume cement paste.*

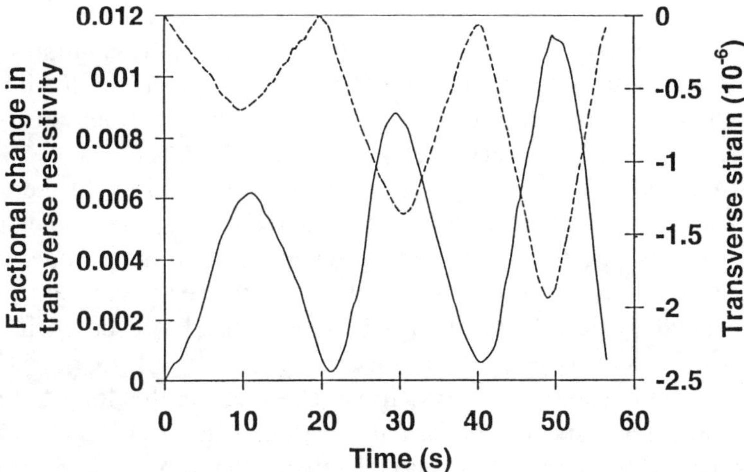

Fig. 3 *Variation of the fractional change in transverse electrical resistivity with time (solid curve) and of the strain with time (dashed curve) during dynamic uniaxial tensile loading at increasing stress amplitudes within the elastic regime for carbon-fiber silica-fume cement paste.*

uniaxial tension. The reversibility of both strain and resistivity is more complete in the longitudinal direction than the transverse direction. The gage factor is 89 and –59 for the longitudinal and transverse resistances respectively.

Figures 4 and 5 show corresponding results for silica-fume cement paste. The strain is essentially totally reversible in both the longitudinal and transverse directions, but the resistivity is only partly reversible in both directions, in contrast to the reversibility of the resistivity when fibers are present (Figs. 2 and 3). As in the case with fibers, both longitudinal and transverse resistivities increase upon uniaxial tension. However, the gage factor is only 7.2 and –7.1 for Figs. 4 and 5 respectively.

Comparison of Figs. 2-3 (with fibers) with Figs. 4-5 (without fibers) shows that fibers greatly enhance the magnitude and reversibility of the resistivity effect. The gage factors are much smaller in magnitude when fibers are absent.

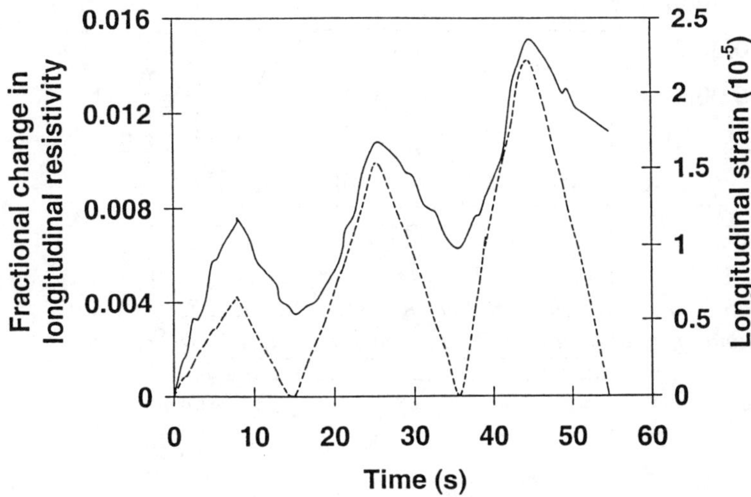

Fig. 4 *Variation of the fractional change in longitudinal electrical resistivity with time (solid curve) and of the strain with time (dashed curve) during dynamic uniaxial tensile loading at increasing stress amplitudes within the elastic regime for silica-fume cement paste.*

The increase in both longitudinal and transverse resistivities upon uniaxial tension for cement pastes, whether with or without fibers, is attributed to defect (e.g., microcrack) generation. In the presence of fibers, fiber bridging across microcracks occurs and slight fiber pull-out occurs upon tension, thus enhancing the possibility of microcrack closing and causing more reversibility in the resistivity change. The fibers are much more electrically conductive than the cement matrix. The presence of the fibers introduces interfaces between fibers and matrix. The degradation of the fiber-matrix interface due to fiber pull-out or other mechanisms is an additional type of defect generation which will increase the resistivity of the composite. Therefore, the presence of fibers greatly increases the gage factor.

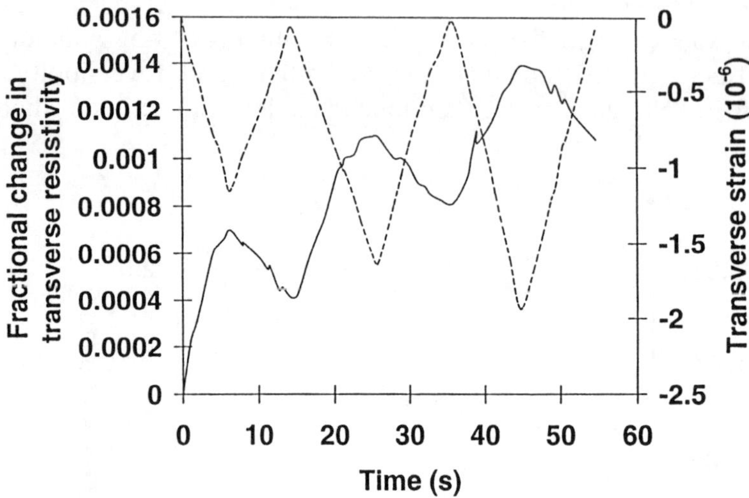

Fig. 5 *Variation of the fractional change in transverse electrical resistivity with time (solid curve) and of the strain with time (dashed curve) during dynamic uniaxial tensile loading at increasing stress amplitudes within the elastic regime for silica-fume cement paste.*

The transverse resistivity increases upon uniaxial tension, even though the Poisson Effect causes the transverse strain to be negative. This means that the effect of the transverse resistivity increase overshadows the effect of the transverse shrinkage. The resistivity increase is a consequence of the uniaxial tension. In contrast, under uniaxial compression, the resistance in the stress direction decreases at 28 days of curing. Hence, the effects of uniaxial tension on the transverse resistivity and of uniaxial compression on the longitudinal resistivity are different; the gage factors are negative and positive for these cases respectively.

The similarity of the resistivity change in longitudinal and transverse directions under uniaxial tension suggests similarity for other directions as well. This means that the resistance can be measured in any direction in order to sense the occurrence of tensile loading. Although the gage factor is comparable in both longitudinal and transverse directions, the fractional change in resistance under uniaxial tension is much higher in the longitudinal direction than the transverse direction. Thus, the use of the longitudinal resistance for practical self-sensing is preferred.

3.1.3 Cement-matrix composites for damage sensing

Concrete, whether with or without admixtures, is capable of sensing major and minor damage – even damage during elastic deformation – due to the electrical resistivity increase that accompanies damage [270, 274, 289-291]. That both strain and damage can be sensed simultaneously through resistance measurement means that the strain/stress condition (during dynamic loading) under which damage occurs can be obtained, thus facilitating damage origin identification. Damage is indicated by a resistance increase, which is larger and less reversible when the stress amplitude is higher. The resistance increase can be a sudden increase during loading. It can also be a gradual shift of the baseline resistance.

Fig. 6(a) [290] shows the fractional change in resistivity along the stress axis as well as the strain during repeated compressive loading at an increasing stress amplitude for plain cement paste at 28 days of curing. Fig. 6(b) shows the corresponding variation of stress and strain during the repeated loading. The strain varies linearly with the stress up to the highest stress amplitude (Fig. 6(b)). The strain returns to zero at the end of each cycle of loading. During the first loading, the resistivity increases due to damage initiation. During the subsequent unloading, the resistivity continues to increase, probably due to opening of the microcracks generated during loading.

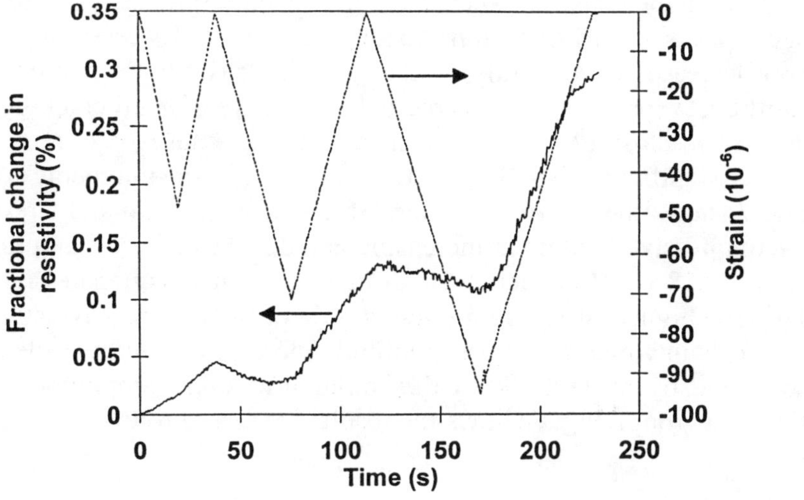

(a)

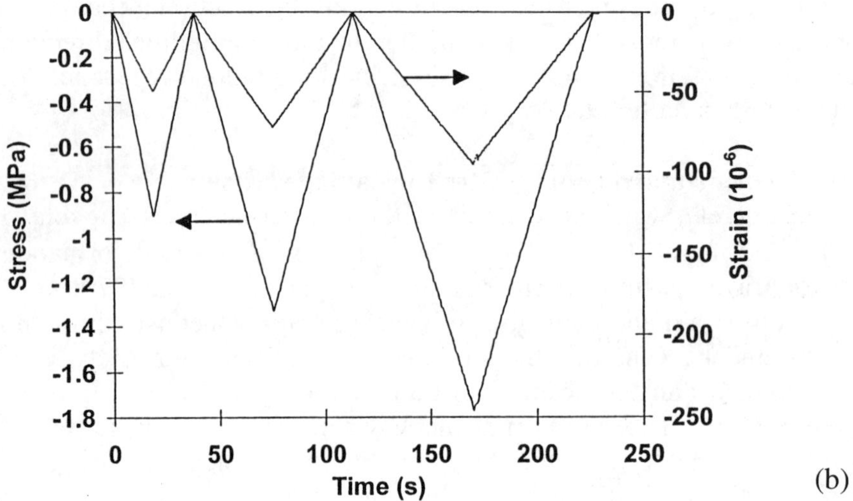

(b)

Fig. 6 *Variation of the fractional change in electrical resistivity with time (a), of the stress with time (b), and of the strain (negative for compressive strain) with time (a,b) during dynamic compressive loading at increasing stress amplitudes within the elastic regime for silica-fume cement paste at 28 days of curing.*

During the second loading, the resistivity decreases slightly as the stress increases up to the maximum stress of the first cycle (probably due to closing of the microcracks) and then increases as the stress increases beyond this value (probably due to the generation of additional microcracks). During unloading in the second cycle, the resistivity increases significantly (probably due to opening of the microcracks). During the third loading, the resistivity essentially does not change (or decreases very slightly) as the stress increases to the maximum stress of the third cycle (probably due to the balance between microcrack generation and microcrack closing). Subsequent unloading causes the resistivity to increase very significantly (probably due to opening of the microcracks).

Fig. 7 shows the fractional change in resistance, strain and stress during repeated compressive loading at increasing and decreasing stress amplitudes for carbon fiber (0.18 vol.%) concrete (with fine and coarse aggregates) at 28 days of curing. The highest stress amplitude is 60% of the compressive strength. A group of cycles in which the stress amplitude increases cycle by cycle and then decreases cycle by cycle back to the initial low stress amplitude is hereby referred to as a group. Fig. 7 shows the results for three groups.

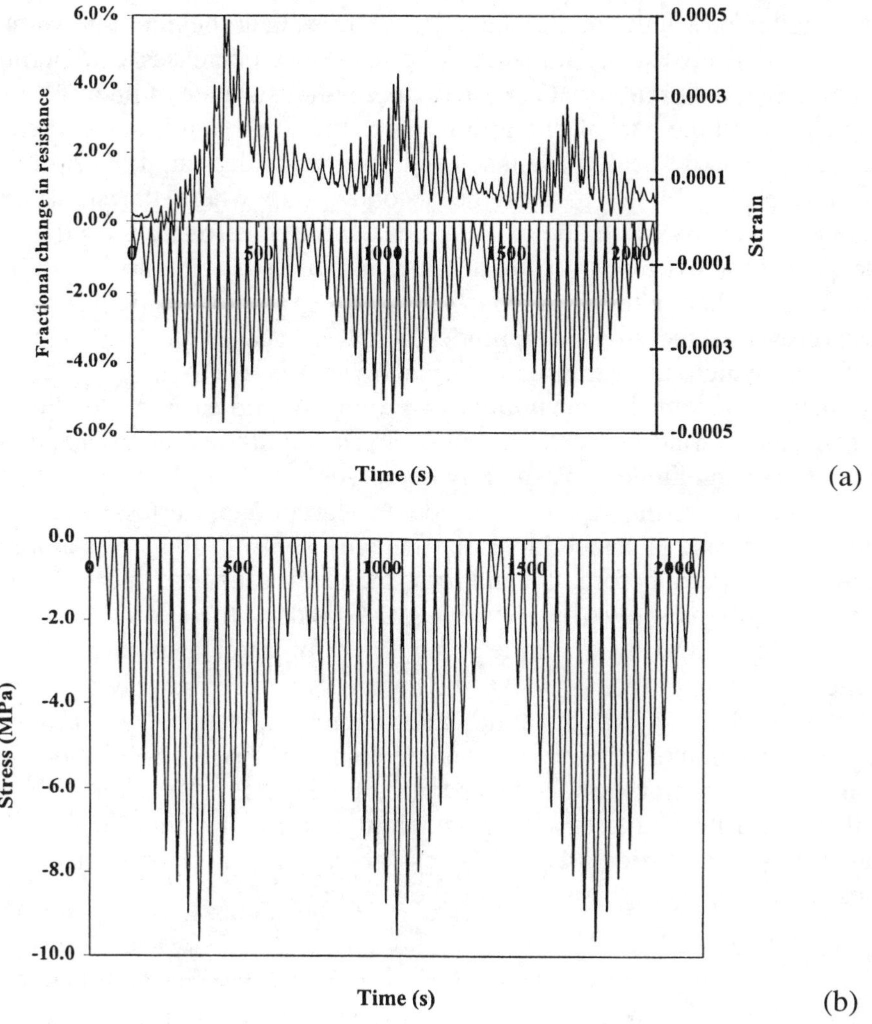

Fig. 7 *Fractional change in resistance (a), strain (a) and stress (b) during repeated compressive loading at increasing and decreasing stress amplitudes, the highest of which was 60% of the compressive strength, for carbon fiber concrete at 28 days of curing.*

The strain returns to zero at the end of each cycle for any of the stress amplitudes, indicating elastic behavior. The resistance decreases upon loading in each cycle, as in Fig. 1. An extra peak at the maximum stress of a cycle grows as the stress amplitude increases, resulting in two peaks per cycle. The original peak (strain induced) occurs at zero stress, while the extra peak (damage induced) occurs at the maximum stress. Hence, during loading from zero stress within a cycle, the resistance drops and then increases sharply, reaching the maximum resistance of the extra peak at the maximum stress of the cycle. Upon subsequent unloading, the resistance decreases and then increases as unloading continues, reaching the maximum resistance of the original peak at zero stress. In the part of this group where the stress amplitude decreases cycle by cycle, the extra peak diminishes and disappears, leaving the original peak as the sole peak. In the part of the second group where the stress amplitude increases cycle by cycle, the original peak (peak at zero stress) is the sole peak, except that the extra peak (peak at the maximum stress) returns in a minor way (more minor than in the first group) as the stress amplitude increases. The extra peak grows as the stress amplitude increases, but, in the part of the second group in which the stress amplitude decreases cycle by cycle, it quickly diminishes and vanishes, as in the first group. Within each group, the amplitude of resistance variation increases as the stress amplitude increases and decreases as the stress amplitude subsequently decreases.

The greater the stress amplitude, the larger and the less reversible is the damage-induced resistance increase (the extra peak). If the stress amplitude has been experienced before, the damage-induced resistance increase (the extra peak) is small, as shown by comparing the result of the second group with that of the first group (Fig. 7), unless the extent of damage is large (Fig. 8 for a highest stress amplitude of >90% the compressive strength). When the damage is extensive (as shown by a modulus decrease), damage-induced resistance increase occurs in every cycle, even at a decreasing stress amplitude, and it can overshadow the strain-induced resistance decrease (Fig. 8). Hence, the damage-induced resistance increase occurs mainly during loading (even within the elastic regime), particularly at a stress above that in prior cycles, unless the stress amplitude is high and/or damage is extensive.

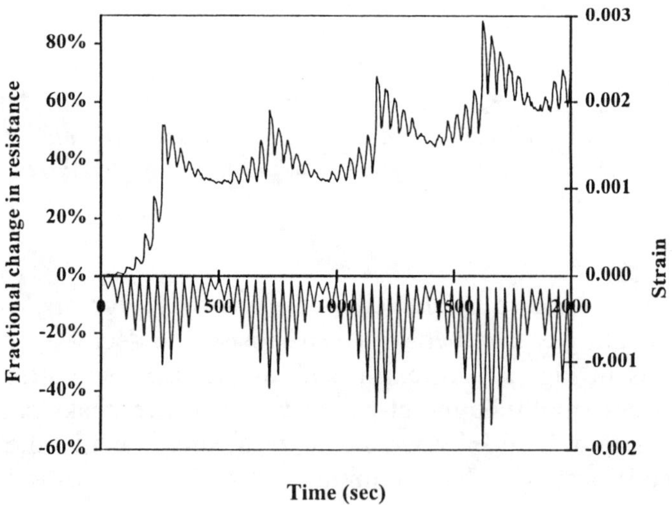

(a)

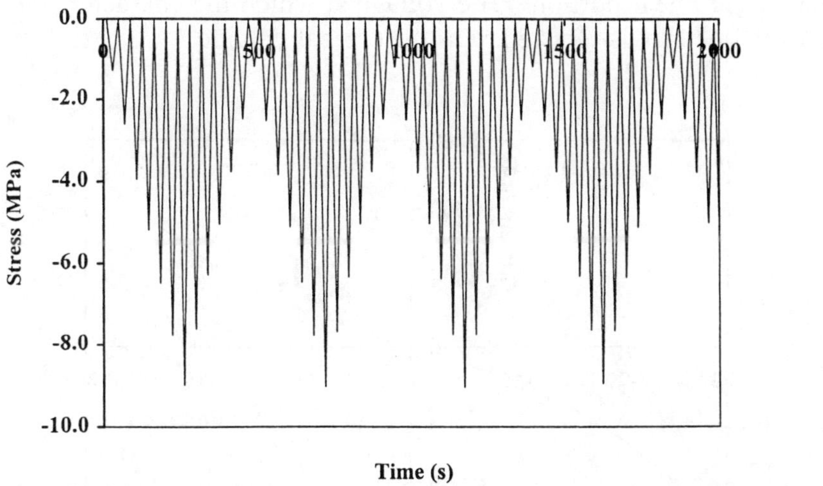

(b)

Fig. 8 *Fractional change in resistance (a), strain (a) and stress (b) during repeated compressive loading at increasing and decreasing stress amplitudes, the highest of which was >90% of the compressive strength, for carbon fiber concrete at 28 days of curing.*

At a high stress amplitude, the damage-induced resistance increase cycle by cycle as the stress amplitude increases causes the baseline resistance to increase irreversibly (Fig. 8). The baseline resistance in the regime of major damage (with a decrease in modulus) provides a measure of the extent of damage (i.e., condition monitoring). This measure works in the loaded or unloaded state. In contrast, the measure using the damage-induced resistance increase (Fig. 7) works only during stress increase and indicates the occurrence of damage (whether minor or major) as well as the extent of damage.

3.1.4 Cement-matrix composites for temperature sensing

A thermistor is a thermometric device consisting of a material (typically a simiconductor, but in this case a cement paste) whose electrical resistivity decreases with rise in temperature. The carbon fiber concrete described in Sec. 3.1.2 for strain sensing is a thermistor due to its resistivity decreasing reversibly with increasing temperature [285]; the sensitivity is comparable to that of semiconductor thermistors. (The effect of temperature will need to be compensated in using the concrete as a strain sensor; Sec. 3.1.2.)

Figure 9 [285] shows the current-voltage characteristic of carbon-fiber (0.5% by weight of cement) silica-fume (15% by weight of cement) cement paste at 38°C during stepped heating. The characteristic is linear below 5 V and deviates positively from linearity beyond 5 V. The resistivity is obtained from the slope of the linear portion. The voltage at which the characteristic starts to deviate from linearity is referred to as the critical voltage.

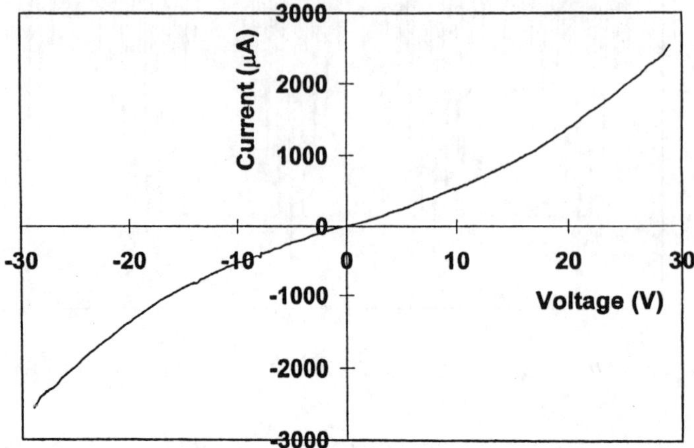

Fig. 9 *Current-voltage characteristic of carbon-fiber silica-fume cement paste at 38°C during stepped heating.*

Figure 10 shows a plot of the resistivity vs. temperature during heating and cooling for carbon-fiber silica-fume cement paste. The resistivity decreases upon heating and the effect is quite reversible upon cooling. That the resistivity is slightly increased after a heating-cooling cycle is probably due to thermal degradation of the material. Fig. 11 shows the Arrhenius plot of log conductivity (conductivity = 1/resistivity) vs. reciprocal absolute temperature. The slope of the plot gives the activation energy, which is 0.390 ± 0.014 and 0.412 ± 0.017 eV during heating and cooling respectively.

Results similar to those of carbon-fiber silica-fume cement paste were obtained with carbon-fiber (0.5% by weight of cement) latex (20% by weight of cement) cement paste, silica-fume cement paste, latex cement paste and plain cement paste. However, for all these four types of cement paste, (i) the resistivity is higher by about an order of magnitude, and (ii) the activation energy is lower by about an order of magnitude, as shown in Table 1. The critical voltage is higher when fibers are absent (Table 1).

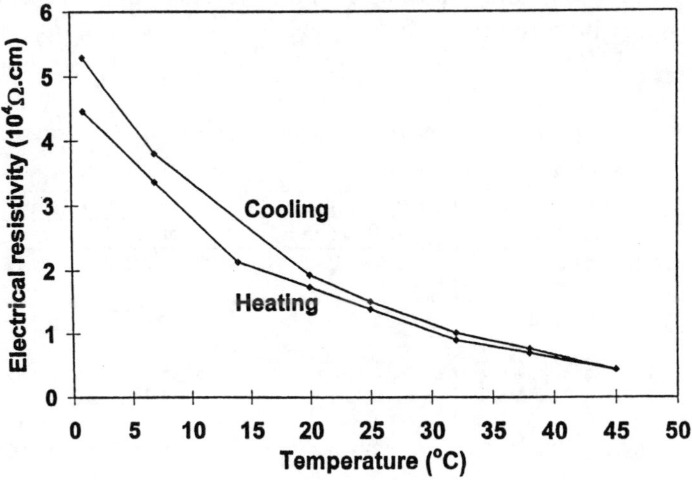

Fig. 10 *Plot of volume electrical resistivity vs. temperature during heating and cooling for carbon-fiber silica-fume cement paste.*

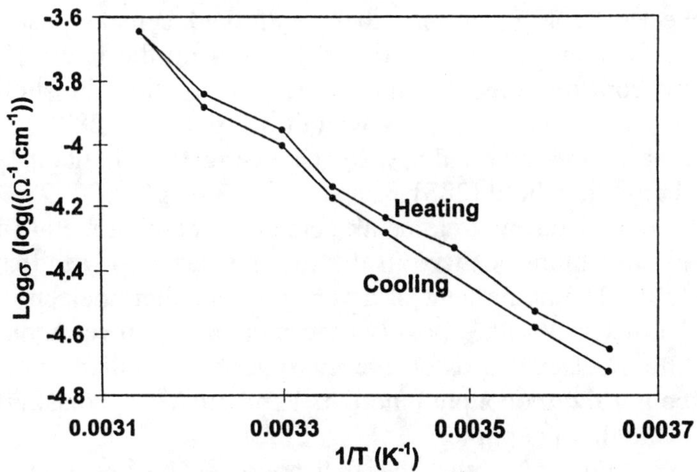

Fig. 11 *Arrhenius plot of log electrical conductivity vs. reciprocal absolute temperature for carbon-fiber silica-fume cement paste.*

The Seebeck [286-288, 320] effect is a thermoelectric effect which is the basis for thermocouples for temperature measurement. This effect involves charge carriers moving from a hot point to a cold point within a material, thereby resulting in a voltage difference between the two points. The Seebeck coefficient is the voltage difference per unit temperature difference between the two points. Negative carriers (electrons) make it more positive and positive carriers (holes) make it more negative.

Table 1. *Resistivity, critical voltage and activation energy of five types of cement paste.*

Formulation	Resistivity at 20°C (Ω.cm)	Critical voltage at 20°C (V)	Activation energy (eV)	
			Heating	Cooling
Plain	(4.87 ± 0.37) x 10^5	10.80 ± 0.45	0.040 ± 0.006	0.122 ± 0.006
Silica fume	(6.12 ± 0.15) x 10^5	11.60 ± 0.37	0.035 ± 0.003	0.084 ± 0.004
Carbon fibers + silica fume	(1.73 ± 0.08) x 10^4	8.15 ± 0.34	0.390 ± 0.014	0.412 ± 0.017
Latex	(6.99 ± 0.12) x 10^5	11.80 ± 0.31	0.017 ± 0.001	0.025 ± 0.002
Carbon fibers + latex	(9.64 ± 0.08) x 10^4	8.76 ± 0.35	0.018 ± 0.001	0.027 ± 0.002

The Seebeck effect in carbon fiber reinforced cement paste involves electrons from the cement matrix [288] and holes from the fibers [286, 287], such that the two contributions are equal at the percolation threshold, a fiber content between 0.5% and 1.0% by weight of cement [288]. The hole contribution increases monotonically with increasing fiber content below and above the percolation threshold [288].

Due to the free electrons in a metal, cement containing metal fibers such as steel fibers is even more positive in the thermoelectric power than cement without fiber [320]. The attainment of a very positive thermoelectric power is attractive, since a material with a positive thermoelectric power and a material with negative thermoelectric power are two very dissimilar materials, the junction of which is a thermocouple junction. (The greater the dissimilarity, the more sensitive is the thermocouple.)

Table 2 and Fig. 12 show the thermopower results. The absolute thermoelectric power is much more positive for all the steel-fiber cement pastes compared to all the carbon-fiber cement pastes. An increase of the steel fiber content from 0.5% to 1.0% by weight of cement increases the absolute thermoelectric power, whether silica fume (or latex) is present or not. An increase of the steel fiber content also increases the reversibility and linearity of the change in Seebeck voltage with the temperature difference between the hot and cold ends, as shown by comparing the values of the Seebeck coefficient obtained during heating and cooling in Table 2. The values obtained during heating and cooling are close for the pastes with the higher steel fiber content, but are not so close for the pastes with the lower steel fiber content. In contrast, for pastes with carbon fibers in place of steel fibers, the change in Seebeck voltage with the temperature difference is highly reversible for both carbon fiber contents of 0.5% and 1.0% by weight of cement, as shown in Table 2 by comparing the values of the Seebeck coefficient obtained during heating and cooling.

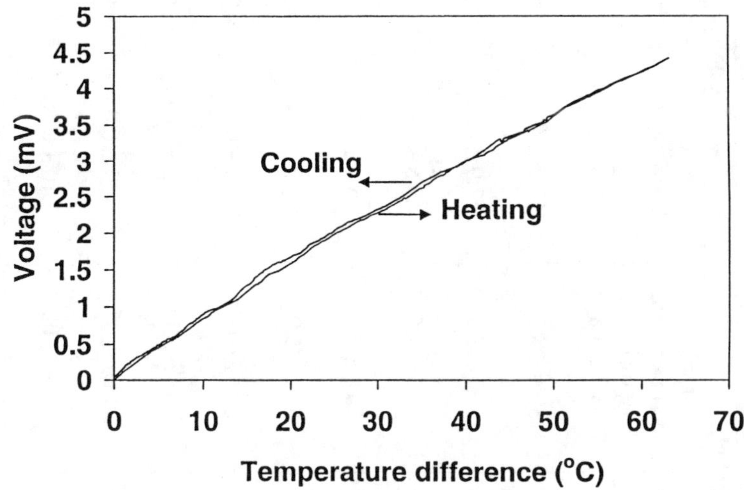

Fig. 12 *Variation of the Seebeck voltage (with copper as the reference) vs. the temperature difference during heating and cooling for steel-fiber silica-fume cement paste containing steel fibers in the amount of 1.0% by weight of cement.*

Table 2 shows that the volume electrical resistivity is much higher for the steel-fiber cement pastes than the corresponding carbon fiber cement pastes. This is attributed to the much lower volume fraction of fibers in the former (Table 2). An increase in the steel or carbon fiber content from 0.5% to 1.0% by weight of cement decreases the resistivity, though the decrease is more significant for the carbon fiber case than the steel fiber case. That the resistivity decrease is not large when the steel fiber content is increased from 0.5% to1.0% by weight of cement and that the resistivity is still high at a steel fiber content of 1.0% by weight of cement suggest that a steel fiber content of 1.0% by weight of cement is below the percolation threshold.

Whether with or without silica fume (or latex), the change of the Seebeck voltage with temperature is more reversible and linear at a steel fiber content of 1.0% by weight of cement than at a steel fiber content of 0.5% by weight of cement. This is attributed to the larger role of the cement matrix at the lower steel fiber content and the contribution of the cement matrix to the irreversibility and non-linearity. Irreversibility and non-linearity are particularly significant when the cement paste contains no fiber.

Table 2. *Volume electical resistivity, Seebeck coefficient (μV/°C) with copper as the reference, and the absolute thermoelectric power (μV/°C) of various cement pastes with steel fibers (S_f) or carbon fibers (C_f).*

Cement paste	Volume fraction fibers	Resistivity (Ω. cm)	Heating		Cooling	
			Seebeck coefficient	Absolute thermoelectric power	Seebeck coefficient	Absolute thermoelectric power
S_f (0.5*)	0.10%	$(7.8 \pm 0.5) \times 10^4$	51.0 ± 4.8	53.3 ± 4.8	45.3 ± 4.4	47.6 ± 4.4
S_f (1.0*)	0.20%	$(4.8 \pm 0.4) \times 10^4$	56.8 ± 5.2	59.1 ± 5.2	53.7 ± 4.9	56.0 ± 4.9
S_f (0.5*) + SF	0.10%	$(5.6 \pm 0.5) \times 10^4$	54.8 ± 3.9	57.1 ± 3.9	52.9 ± 4.1	55.2 ± 4.1
S_f (1.0*) + SF	0.20%	$(3.2 \pm 0.3) \times 10^4$	66.2 ± 4.5	68.5 ± 4.5	65.6 ± 4.4	67.9 ± 4.4
S_f (0.5*) + L	0.085%	$(1.4 \pm 0.1) \times 10^5$	48.1 ± 3.2	50.4 ± 3.2	45.4 ± 2.9	47.7 ± 2.9
S_f (1.0*) + L	0.17%	$(1.1 \pm 0.1) \times 10^5$	55.4 ± 5.0	57.7 ± 5.0	54.2 ± 4.5	56.5 ± 4.5
C_f (0.5*) + SF	0.48%	$(1.5 \pm 0.1) \times 10^4$	-1.45 ± 0.09	0.89 ± 0.09	-1.45 ± 0.09	0.89 ± 0.09
C_f (1.0*) + SF	0.95%	$(8.3 \pm 0.5) \times 10^2$	-2.82 ± 0.11	-0.48 ± 0.11	-2.82 ± 0.11	-0.48 ± 0.11
C_f (0.5*) + L	0.41%	$(9.7 \pm 0.6) \times 10^4$	-1.20 ± 0.05	1.14 ± 0.05	-1.20 ± 0.05	1.14 ± 0.05
C_f (1.0*) + L	0.82%	$(1.8 \pm 0.2) \times 10^3$	-2.10 ± 0.08	0.24 ± 0.08	-2.10 ± 0.08	0.24 ± 0.08

* % by weight of cement

SF: silica fume

L: latex

From the practical point of view, the steel-fiber silica-fume cement paste containing steel fibers in the amount of 1.0% by weight of cement is particularly attractive for use in temperature sensing, as the absolute thermoelectric power is the highest (68 μV/°C) and the variation of the Seebeck voltage with the temperature difference between the hot and cold ends is reversible and linear. The absolute thermoelectric power is as high as those of commercial thermocouple materials.

Joints between concretes with different values of the thermoelectric power, as made by multiple pouring, provide concrete thermocouples [321].

3.1.5 Cement-matrix composites for reflecting electromagnetic radiation

Cement-matrix composites containing 0.1 μm diameter discontinuous carbon filaments are effective for reflecting radio waves [224]. Due to the skin effect, conventional carbon fibers (7-15 μm diameter) are much less effective. The reflectivity renders the ability to shield EMI and to provide lateral guidance in the automatic highway technology [224].

3.2 Polymer-matrix composites for smart structures

Polymer-matrix composites for structural applications typically contain continuous fibers such as carbon, polymer and glass fibers, as continuous fibers tend to be more effective than short fibers as a reinforcement. Polymer-matrix composites with continuous carbon fibers are used for aerospace, automobile and civil structures. (In contrast, continuous fibers are too expensive for reinforcing concrete.) Due to the fact that carbon fibers are electrically conducting, whereas polymer and glass fibers are not, carbon fiber composites are predominant among polymer-matrix composites that are intrinsically smart.

3.2.1 Background on polymer-matrix structural composites

Polymer-matrix composites containing continuous carbon fibers are important structural materials due to their high tensile strength, high tensile modulus and low density. They are used for lightweight structures such as satellites, aircraft, automobiles, bicycles, ships, submarines, sporting goods, wheel chairs, armor and rotating machinery (such as turbine blades and helicoptor rotors). Due to the recent emphasis on repair of civil infrastructural systems, composites are beginning to be used for the repair of concrete structures and for bridges, even though they are much more expensive than concrete. As the price of carbon fibers has been dropping steadily during the last two decades, the spectrum of applications has been widening tremendously.

The continuous carbon fibers used are primarily either based on polyacrylonitrile (PAN) or mesophase pitch. Mesophase-pitch-based carbon fibers, if heat treated to high temperatures exceeding 2500°C, can be graphitized and attain very high values of the tensile modulus and thermal conductivity (in-plane), in addition to improved oxidation resistance. The high thermal conductivity is attractive for thermal management, which is particularly important for electronics (i.e., heat sinks, etc.). However, the graphitezed fibers tend to be relatively low in strength, due to the ease of shear between the graphite layers, and they are very expensive. On the other hand, PAN-based fibers cannot be graphitized, though they compete well with mesophase-pitch-based fibers which have not been graphitized, in that both materials exhibit reasonably high values of both strength and modulus and are not very expensive. These fibers are the most widely used among carbon fibers. The fabrication of both pitch-based and PAN-based carbon fibers involves stabilization (infusibilization) and then carbonization (conversion from hydrocarbon molecules to a carbon network). Graphitization optionally follows carbonization.

Due to the importance of carbon fiber polymer-matrix composites for structural applications, much investigation has been made on the mechanical behavior of these materials. Much less work has been done to study the electrical behavior [322-329]. On the other hand, due to the fact that carbon fibers are much more conducting than the polymer matrix, the electrical behavior gives much information on the microstructure, such as the degree of fiber alignment, the number of fiber-fiber contacts, the amount of delamination and the extent of fiber breakage. Such information is not only useful for scientific understanding of the properties of the composite, but is also valuable for the purpose of rendering the composite the ability to sense its strain, damage and temperature in real time via electrical measurement. In other words, the strain, damage and temperature affect the electrical behavior, such as the electrical resistance, which thus serves to indicate strain, damage and temperature. In this way, the composite is self-sensing, i.e., intrinsically smart, without the need for attached or embedded sensors (such as optical fibers, acoustic sensors and piezoelectric sensors), which raise the cost, reduce the durability and, in the case of embedded sensors, weaken the structure.

Carbon fibers are electrically conducting, while the polymer matrix is electrically insulating (except for the rare situation in which the polymer is an electrically conducting one [330]). The continuous fibers in a composite laminate are in the form of layers called laminae. Each lamina comprises many bundles (called tows) of fibers in a polymer matrix. Each tow consists of thousands of fibers. There may or may not be twist in a tow. Each fiber has a diameter typically ranging from 7 to 12 μm. The tows within a lamina are typically oriented in the same direction, but tows in different laminae may or may not be in the same direction. A laminate with tows in all the laminae oriented in the same direction is said to unidirectional. A laminate with tows in adjacent laminae oriented at an angle of 90° is said to be crossply. In general, an angle of 45° and other angles may also be involved for the various laminae, as desired for attaining the mechanical properties required for thc laminate in various directions in the plane of the laminate.

Within a lamina with tows in the same direction, the electrical conductivity is highest in the fiber direction. In the transverse direction in the plane of the lamina, the conductivity is not zero, even though the polymer matrix is insulating. This is because there are contacts between fibers of adjacent tows [331]. In other words, a fraction of the fibers of one tow touch a fraction of the fiber of an adjacent tow here and there along the length of the fibers. These contacts result from the fact that fibers are not perfectly straight or parallel (even though the lamina is said to be unidirectional), and that the flow of the polymer matrix (or resin) during composite fabrication can cause a fiber to be not completely covered by the polymer or resin (even though, prior to composite fabrication, each fiber may be completely covered by the polymer or resin, as in the case of a prepreg, i.e., a fiber sheet impregnated with the polymer or resin). Fiber waviness is known as marcelling. Thus, the transverse conductivity gives information on the number of fiber-fiber contacts in the plane of the lamina.

For similar reasons, the contacts between fibers of adjacent laminae cause the conductivity in the through-thickness direction (direction perpendicular to the plane of the laminate) to be non-zero. Thus, the through-thickness conductivity gives information on the number of fiber-fiber contacts between adjacent laminae.

Matrix cracking between the tows of a lamina decreases the number of fiber-fiber contacts in the plane of the lamina, thus decreasing the transverse conductivity. Similarly, matrix cracking between adjacent laminae (as in delamination [332]) decreases the number of fiber-fiber contacts between adjacent laminae, thus decreasing the through-thickness conductivity. This means that the transverse and through-thickness conductivities can indicate damage in the form of matrix cracking.

Fiber damage (as distinct from fiber fracture) decreases the conductivity of a fiber, thereby decreasing the longitudinal conductivity (conductivity in the fiber direction). However, due to the brittleness of carbon fibers, the decrease in conductivity due to fiber damage prior to fiber fracture is rather small [333].

Fiber fracture causes a much larger decrease in the longitudinal conductivity of a lamina than fiber damage. If there is only one fiber, a broken fiber results in an open circuit, i.e., zero conductivity. However, a lamina has a large number of fibers and adjacent fibers can make contact here and there. Therefore, the portions of a broken fiber still contribute to the longitudinal conductivity of the lamina. As a result, the decrease in conductivity due to fiber fracture is less than what it would be if a broken fiber did not contribute to the conductivity. Nevertheless, the effect of fiber fracture on the longitudinal conductivity is significant, so that the longitudinal conductivity can indicate damage in the form of fiber fracture [334].

The through-thickness volume resistance of a laminate is the sum of the volume resistance of each of the laminae in the through-thickness direction and the contact resistance of each of the interfaces between adjacent laminae (i.e., the interlaminar interface). For example, a laminate with eight laminae has eight volume resistances and seven contact resistances, all in the through-thickness direction. Thus, to study the interlaminar interface, it is better to measure the contact resistance between two laminae rather than the through-thickness volume resistance of the entire laminate.

Measurement of the contact resistance between laminae can be made by allowing two laminae (strips) to contact at a junction and using the two ends of each strip for making four electrical contacts [335]. An end of the top strip and an end of the bottom strip serve as contacts for passing current. The other end of the top strip and the other end of the bottom strip serve as contacts for voltage measurement. The fibers in the two strips can be in the same direction or in different directions. This method is a form of the four-probe method of electrical resistance. The configuration is illustrated in Fig. 13 for crossply and unidirectional laminates. To make sure that the volume resistance within a lamina in the through-thickness direction does not contribute to the measured resistance, the fibers at each end of a lamina strip should be electrically shorted together by using silver paint or other conducting media. The measured resistance is the contact resistance of the junction. This resistance, multiplied by the area of the junction, gives the contact resistivity, which is independent of the area of the junction and just depends on the nature of the interlaminar interface. The unit of the contact resistivity is $\Omega.m^2$, whereas that of the volume resistivity is $\Omega.m$.

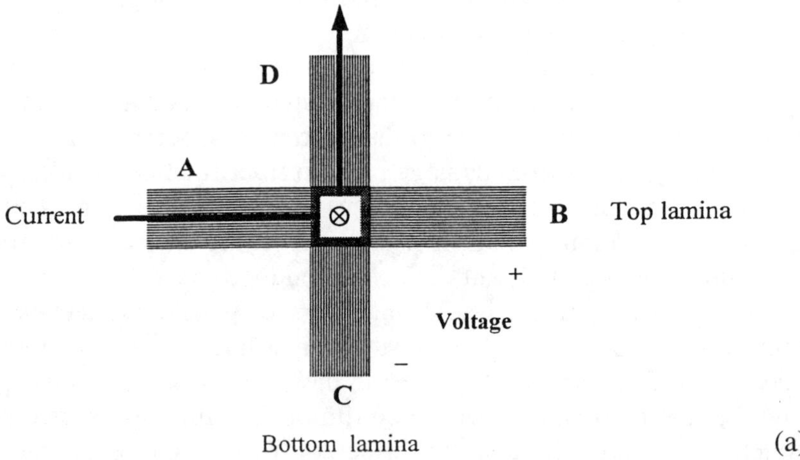

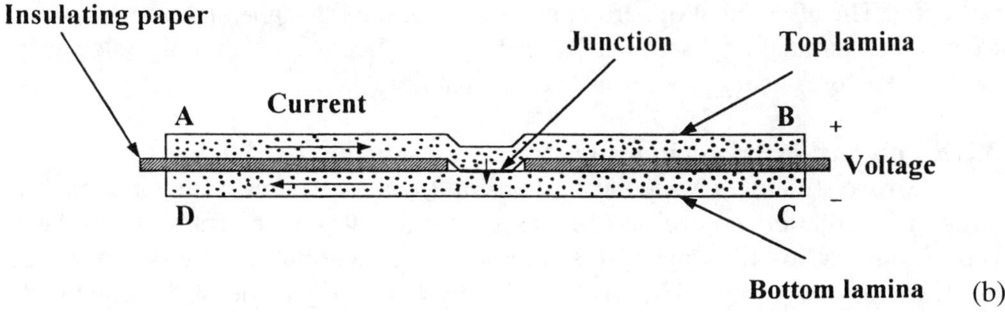

Fig. 13 *Specimen configuration for measurement of the contact electrical resistivity between laminae. (a) Crossply laminae. (b) Unidirectional laminae.*

The structure of the interlaminar interface tends to be more prone to change than the structure within a lamina. For example, damage in the form of delamination is much more common than damage in the form of fiber fracture. Moreover, the structure of the interlaminar interface is affected by the interlaminar stress (whether thermal stress or curing stress), which is particularly significant when the laminae are not unidirectional (as the anisotropy within each lamina enhances the interlaminar stress). The structure of the interlaminar interface also depends on the extent of consolidation of the laminae during composite fabrication. The contact resistance provides a sensitive probe of the structure of the interlaminar interface.

The measurement of the volume resistivity in the through-thickness direction can be conducted by using the four-probe method, in which each of the two current contacts is in the form of a conductor loop (made by silver paint, for example) on each of the two outer surfaces of the laminate in the plane of the laminate and each of the two voltage contacts is in the form of a conductor dot within the loop [332]. An alternate method is to have four of the laminae in the laminate be extra long so as to extend out for the purpose of serving as electrical leads [336]. The two outer leads are for current contacts, the two inner leads are for voltage contacts. The use of a thin metal wire inserted at an end into the interlaminar space during composite fabrication in order to serve as an electrical contact is not recommended, because the quality of the electrical contact between the metal wire and carbon fibers is hard to control and the wire is intrusive to the composite. The alternate method is less convenient than the method involving loops and dots, but it approaches more closely the ideal four-probe method.

In order to attain zero conductivity in the through-thickness direction of a laminate, it is necessary to use an insulating layer between two adjacent laminae [337]. The insulating layer can be a piece of writing paper. Tissue paper is ineffective in preventing contacts between fibers of adjacent laminae, due to its porosity. The attainment of zero conductivity in the through-thickness direction allows the laminate to serve as a capacitor. This means that the structural composite stores energy by serving as a capacitor.

3.2.2 Polymer-matrix composites for strain sensing

Smart structures which can monitor their own strain are valuable for structural vibration control. Self-monitoring of strain (reversible) has been achieved in carbon fiber epoxy-matrix composites without the use of embedded or attached sensor [336, 338-341], as the electrical resistance of the composite in the through-thickness or longitudinal direction changes reversibly with longitudinal strain (gage factor up to 40) due to change in the degree of fiber alignment. Tension in the fiber direction of the composite increases the degree of fiber alignment, thereby decreasing the chance for fibers of adjacent laminae to touch one another. As a consequence, the through-thickness resistance increases while the longitudinal resistance decreases.

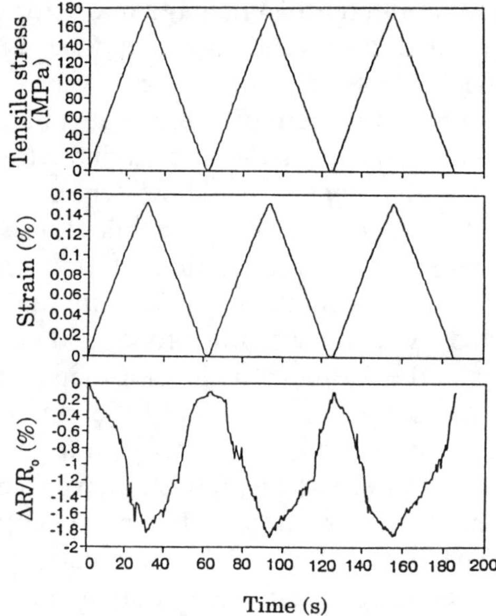

Fig. 14 *Longitudinal stress and strain and fractional resistance increase ($\Delta R/R_0$) obtained simultaneously during cyclic tension at a stress amplitude equal to 14% of the breaking stress for continuous fiber epoxy-matrix composite.*

Fig. 14 [339] shows the change in longitudinal resistance during cyclic longitudinal tension in the elastic regime for a unidirectional continuous carbon fiber epoxy-matrix composite with eight fiber layers (laminae). The stress amplitude is equal to 14% of the breaking stress. The strain returns to zero at the end of each cycle. Because of the small strains involved, the fractional resistance change $\Delta R/R_0$ is essentially equal to the fractional change in resistivity. The longitudinal $\Delta R/R_0$ decreases upon loading and increases upon unloading in every cycle, such that R irreversibly decreases slightly after the first cycle (i.e., $\Delta R/R_0$ does not return to 0 at the end of the first cycle). At higher stress amplitudes, the effect is similar, except that both the reversible and irreversible parts of $\Delta R/R_0$ are larger.

Fig. 15 [339] shows the change in the through-thickness resistance during cyclic longitudinal tension in the elastic regime for the same composite. The stress amplitude is equal to 14% of the breaking stress. The through-thickness $\Delta R/R_0$ increases upon loading and decreases upon unloading in every cycle, such that R irreversibly decreases slightly after the first cycle (i.e., $\Delta R/R_0$ does not return to 0 at the end of the first cycle). Upon increasing the stress amplitude, the effect is similar, except that the reversible part of $\Delta R/R_0$ is larger.

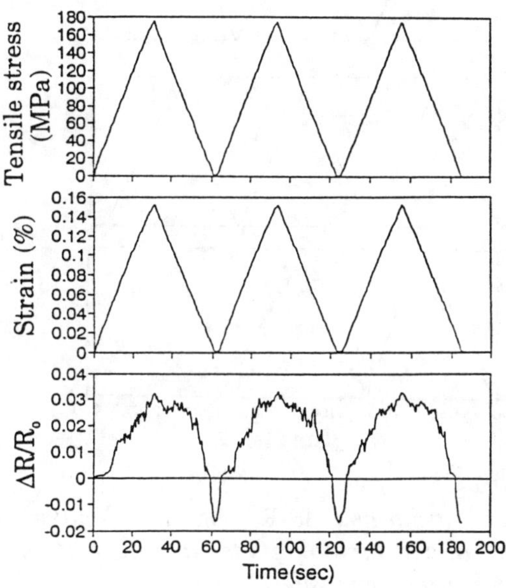

Fig. 15 *Longitudinal stress and strain and the through-thickness* $\Delta R/R_0$ *obtained simultaneously during cyclic tension at a stress amplitude equal to 14% of the breaking stress for continuous fiber epoxy-matrix composite.*

The strain sensitivity (gage factor) is defined as the reversible part of $\Delta R/R_0$ divided by the longitudinal strain amplitude. It is negative (from –18 to –12) for the longitudinal $\Delta R/R_0$ and positive (17-24) for the through-thickness

$\Delta R/R_0$. The magnitudes are comparable for the longitudinal and through-thickness strain sensitivities. As a result, whether the longitudinal R or the through-thickness R is preferred for strain sensing just depends on the convenience of electrical contact application for the geometry of the particular smart structure.

Fig. 16 [339] shows the compressive stress, strain and longitudinal $\Delta R/R_0$ obtained simultaneously during cyclic compression at stress amplitudes equal to 14% of the breaking stress for a similar composite having 24 rather than 8 fiber layers. The longitudinal $\Delta R/R_0$ increases upon compressive loading and decreases upon unloading in every cycle, such that resistance R irreversibly increases very slightly after the first cycle. The magnitude of the gage factor is lower in compression (-1.2) than in tension (from –18 to –12).

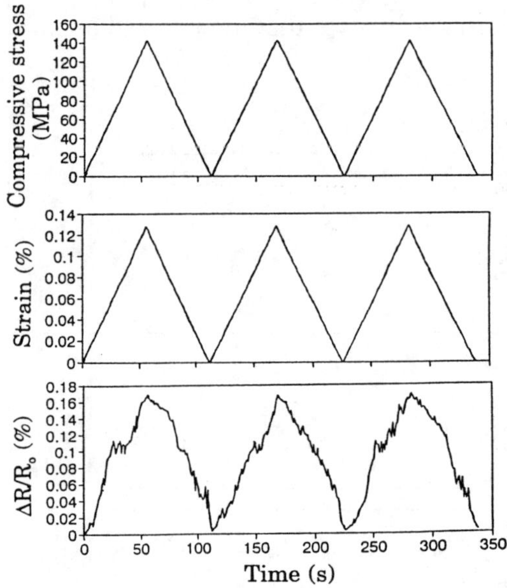

Fig. 16 *Longitudinal stress, strain and* $\Delta R/R_0$ *obtained simultaneously during cyclic compression (longitudinal) at a stress amplitude equal to 14% of the breaking stress for continuous fiber epoxy-matrix composite.*

A dimensional change without any resistivity change would have caused longitudinal R to increase during tensile loading and decrease during compressive loading. In contrast, the longitudinal R decreases upon tensile loading and increases upon compressive loading. In particular, the magnitude of $\Delta R/R_0$ under tension is 7-11 times that of $\Delta R/R_0$ calculated by assuming that $\Delta R/R_0$ is only due to dimensional change and not due to any resistivity change.

Hence the contribution of $\Delta R/R_0$ from the dimensional change is negligible compared to that from the resistivity change.

The irreversible behavior, though small compared to the reversible behavior, is such that R (longitudinal or through-thickness) under tension is irreversibly decreased after the first cycle. This behavior is attributed to the irreversible disturbance to the fiber arrangement at the end of the first cycle, such that the fiber arrangement becomes less neat. A less neat fiber arrangement means more chance for the adjacent fiber layers to touch one another.

3.2.3 Polymer-matrix composites for damage sensing

Self-monitoring of damage (whether due to stress or temperature, under static or dynamic conditions) has been achieved in continuous carbon fiber polymer-matrix composites, as the electrical resistance of the composite changes with damage [332, 338, 342-355]. Minor damage in the form of slight matrix damage and/or disturbance to the fiber arrangement is indicated by the longitudinal and through-thickness resistance decreasing irreversibly due to increase in the number of contacts between fibers, as shown after one loading cycle in Fig. 14-16. More significant damage in the form of delamination or interlaminar interface degradation is indicated by the through-thickness resistance (or, more exactly, the contact resistivity of the interlaminar interface) increasing due to decrease in the number of contacts between fibers of different laminae. Major damage in the form of fiber breakage is indicated by the longitudinal resistance increasing irreversibly.

During mechanical fatigue, delamination was observed to begin at 30% of the fatigue life, whereas fiber breakage was observed to begin at 50% of the fatigue life. Fig. 17 [342] shows an irreversible resistance increase occurring at about 50% of the fatigue life during tension-tension fatigue testing of a unidirectional continuous carbon fiber epoxy-matrix composite. The resistance and stress are in the fiber direction. The reversible changes in resistance are due to strain, which causes the resistance to decrease reversibly in each cycle, as in Fig. 14.

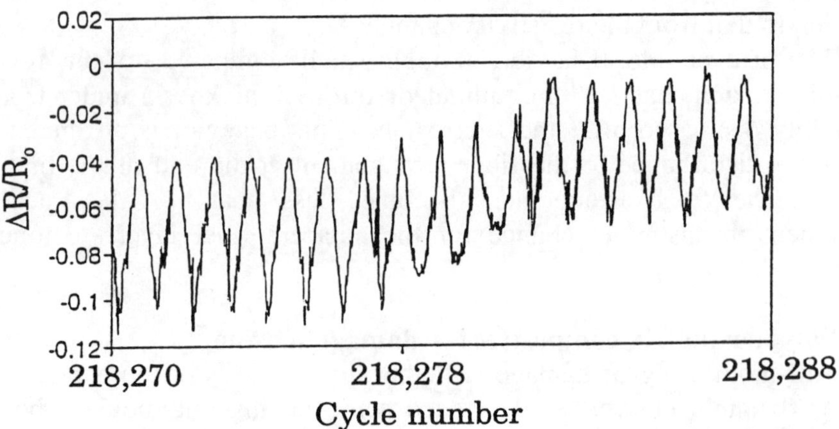

Fig. 17 *Variation of longitudinal* $\Delta R/R_0$ *with cycle number during tension-tension fatigue testing for a carbon fiber epoxy-matrix composite. Each cycle of reversible decrease in resistance is due to strain. The irreversible increase in resistance at around cycle no. 218,281 is due to damage in the form of fiber breakage.*

Fig. 18 shows the variation of the contact resistivity with temperature during thermal cycling. The temperature is repeatedly increased to various levels. A group of cycles in which the temperature amplitude increases cycle by cycle and then decreases cycle by cycle back to the initial low temperature amplitude is referred to as a group.

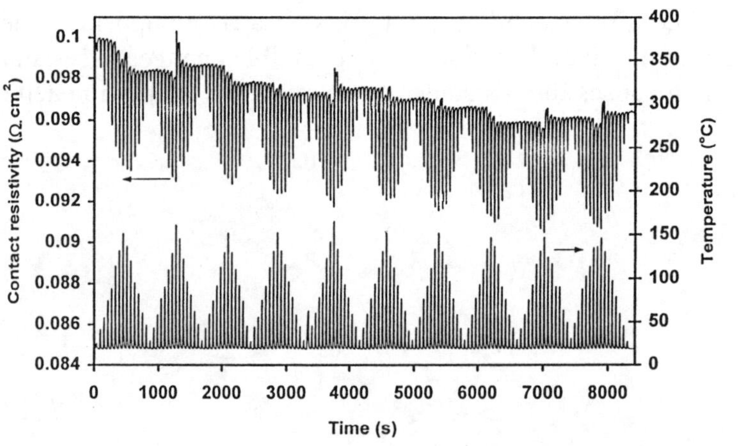

(a)

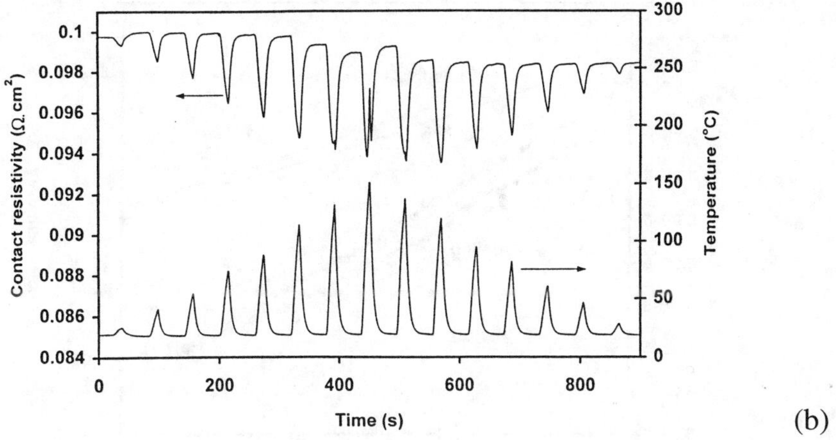

(b)

Fig. 18 *Variation of contact electrical resistivity with time and of the temperature with time during thermal cycling of a carbon fiber epoxy-matrix composite. (b) is the magnified view of the first 900 s of (a).*

Fig. 18(a) shows the results of the first 10 groups, while Fig. 18(b) shows the first group only. The contact resistivity decreases upon heating in every cycle of every group. At the highest temperature (150°C) of a group, a spike of resistivity increase occurs, as shown in Fig. 18(b). It is attributed to damage at the interlaminar interface. In addition, the baseline resistivity (i.e., the top envelope) gradually and irreversibly shifts downward as cycling progresses, as shown in Fig. 18(a). The baseline decrease is probably due to matrix damage within a lamina and the resulting decrease in modulus and hence decrease in residual stress.

3.2.4 Polymer-matrix composites for temperature sensing

Continuous carbon fiber epoxy-matrix composites provide temperature sensing by serving as thermistors [335, 356] and thermocouples [357].

The thermistor function stems from the reversible decrease of the contact electrical resistivity at the interface between fiber layers (laminae) on temperature. Fig. 19 shows the variation of the contact resistivity ρ_c with temperature during reheating and subsequent cooling, both at 0.15°C/min, for carbon fiber epoxy-matrix composites cured at 0 and 0.33 MPa. The corresponding Arrhenius plots of log contact conductivity (inverse of contact resistivity) versus inverse absolute temperature during heating are shown in Fig. 20.

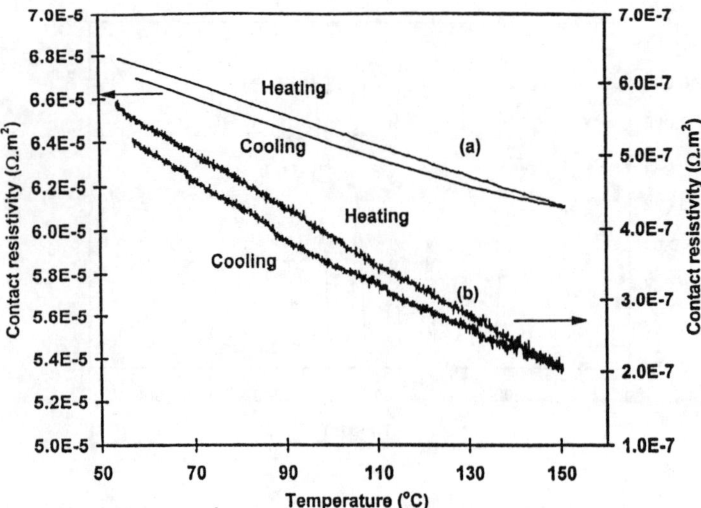

Fig. 19 *Variation of contact electrical resistivity with temperature during heating and cooling of carbon fiber epoxy-matrix composites at 0.15 °C/min (a) for composite made without any curing pressure and (b) for composite made with a curing pressure 0.33 MPa.*

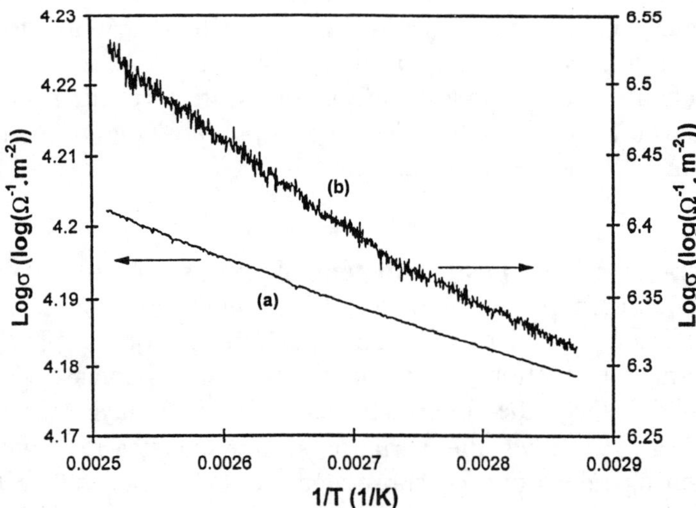

Fig. 20 *Arrhenius plot of log contact conductivity vs. inverse absolute temperature during heating of carbon fiber epoxy-matrix composites at 0.15 °C/min (a) for composite made without any curing pressure and (b) for composite made with curing pressure 0.33 MPa.*

From the slope (negative) of the Arrhenius plot, which is quite linear, the activation energy can be calculated. The linearity of the Arrhenius plot means that the activation energy does not change throughout the temperature variation. This activation energy is the energy for electron jumping from one lamina to the other. Electronic excitation across this energy enables conduction in the through-thickness direction.

The activation energies, thicknesses and room temperature contact resistivities for samples made at different curing pressures and composite configurations are shown in Table 3.

Table 3. *Activation energy for various carbon fiber epoxy-matrix composites. The standard deviations are shown in parentheses.*

Composite Configuration	Curing pressure (MPa)	Composite thickness (mm)	Contact Resistivity ρ_{co} ($\Omega.m^2$)	Activation energy (kJ/mol)		
				Heating at 0.15°C/min	Heating at 1°C/min	Cooling at 0.15°C/min
Crossply	0	0.36	7.3×10^{-5}	1.26 (2×10^{-3})	1.24 (3×10^{-3})	1.21 (8×10^{-4})
	0.062	0.32	1.4×10^{-5}	1.26 (4×10^{-3})	1.23 (7×10^{-3})	1.23 (4×10^{-3})
	0.13	0.31	1.8×10^{-5}	1.62 (3×10^{-3})	1.57 (4×10^{-3})	1.55 (2×10^{-3})
	0.19	0.29	5.4×10^{-6}	2.14 (3×10^{-3})	2.15 (3×10^{-3})	2.13 (1×10^{-3})
	0.33	0.26	4.0×10^{-7}	11.4 (4×10^{-2})	12.4 (8×10^{-2})	11.3 (3×10^{-2})
Unidirectional	0.42	0.23	2.9×10^{-5}	1.02 (3×10^{-3})	0.82 (4×10^{-3})	0.78 (2×10^{-3})

All the activation energies were calculated based on the data at 75 – 125°C. In this temperature regime, the temperature change was very linear and well controlled. From Table 3 it can be seen that, for the same composite configuration (crossply), the higher the curing pressure, the smaller the composite thickness (because of more epoxy being squeezed out), the lower the contact resistivity, and the higher was the activation energy. A smaller composite thickness corresponds to a higher fiber volume fraction in the composite. During curing and subsequent cooling, the matrix shrinks, so a longitudinal compressive stress will develop in the fibers. For carbon fibers, the modulus in the longitudinal direction is much higher than that in the transverse direction. Moreover, the carbon fibers are continuous in the longitudinal direction. Thus, the overall shrinkage in the longitudinal direction tends to be less than that in the transverse direction. Therefore, there will be a residual interlaminar stress in the two crossply layers in a given direction. This stress accentuates the barrier for the electrons to jump from one lamina to the other. After curing and subsequent cooling, heating will decrease the thermal stress. Both the thermal stress and the curing stress contribute to the residual interlaminar stress. Therefore, the higher the curing pressure, the larger the fiber volume fraction, the greater the residual interlaminar stress, and the higher was the activation energy, as shown in Table 3. Besides the residual stress, thermal expansion can also affect contact resistance by changing the contact area. However, calculation shows that the contribution of thermal expansion is less than one-tenth of the observed change in contact resistance with temperature.

The electron jump primarily occurs at points where direct contact occurs between fibers of the adjacent laminae. The direct contact is possible due to the flow of thc cpoxy rcsin during compositc fabrication and duc to thc slight waviness of the fibers, as explained in Ref. 332 in relation to the through-thickness volume resistivity of a carbon fiber epoxy-matrix composite.

The curing pressure for the sample in the unidirectional composite configuration was higher than that of any of the crossply samples (Table 3). Consequently, the thickness was the lowest. As a result, the fiber volume fraction was the highest. However, the contact resistivity of the unidirectional sample was the second highest rather than being the lowest, and its activation energy was the lowest rather than the highest. The low activation energy is consistent with the fact that there was no CTE or curing shrinkage mismatch between the two unidirectional laminae and, as a result, no interlaminar stress between the laminae. This low value supports the notion that the interlaminar stress is important in affecting the activation energy. The high contact resistivity for the unidirectional case can be explained in the following way. In the crossply samples, the pressure during curing forced the fibers of the two laminae to press on to one another and hence contact tightly. In the unidirectional sample, the fibers of one of the laminae just sank into the other lamina at the junction, so pressure helped relatively little in the contact between fibers of adjacent laminae. Moreover, in the crossply situation, every fiber at the lamina-lamina interface contacted many fibers of the other lamina, while, in the unidirectional situation, every fiber had little chance to contact the fibers of the other lamina. Therefore, the number of contact points between the two laminae was less for the unidirectional sample than the crossply samples.

The thermocouple function stems from the use of n-type and p-type carbon fibers (as obtained by intercalation) in different laminae. The thermocouple sensitivity and linearity are as good as or better than those of commercial thermocouples. By using two laminae that are crossply, a two-dimensional array of thermistors or thermocouple junctions are obtained, thus allowing temperature distribution sensing.

Table 4. *Seebeck coefficient (μV/°C) and absolute thermoelectric power (μV/°C) of carbon fibers and thermocouple sensitivity (μV/°C) of epoxy-matrix composite junctions. All junctions are unidirectional unless specified as crossply. The temperature range is 20-110°C.*

	Seebeck coefficient with copper as the reference (μV/°C)	Absolute thermoelectric power (μV/°C)	Thermocouple sensitivity (μv/°c)
P-25*	- 0.8	+ 1.5	
T-300*	+ 5.0	+ 7.3	
P-25* + T-300*			+ 5.5
P-25* + T-300* (crossply)			+ 5.4
P-100*	+ 1.7	+ 4.0	
P-120*	+ 3.2	+ 5.5	
P-100 (Na)	+ 48	+ 50	
P-100 (Br_2)	- 43	- 41	
P-100 (Br_2) + P-100 (Na)			+ 82
P-120 (Na)	+ 42	+ 44	
P-120 (Br_2)	- 38	- 36	
P-120 (Br_2) + P-120 (Na)			+ 74

*Pristine (i.e., not intercalated)

Table 4 shows the Seebeck coefficient and the absolute thermoelectric power of carbon fibers and the thermocouple sensitivity of epoxy-matrix composite junctions. A negative value of the absolute thermoelectric power indicates p-type behavior; a positive value indicates n-type behavior. Pristine P-25 is slightly n-type; pristine T-300 is strongly n-type. A junction comprising pristine P-25 and pristine T-300 has a positive thermocouple sensitivity that is close to the difference of the Seebeck coefficients (or the absolute thermoelectric powers) of T-300 and P-25, whether the junction is unidirectional or crossply. Pristine P-100 and pristine P-120 are both slightly n-type. Intercalation with sodium causes P-100 and P-120 to become strongly n-type. Intercalation with bromine causes P-100 and P-120 to become strongly p-type. A junction comprising bromine intercalated P-100 and sodium intercalated P-100 has a positive thermocouple sensitivity that is close to the sum of the magnitudes of the absolute thermoelectric powers of the bromine intercalated P-100 and the sodium intercalated P-100. Similarly, a junction comprising bromine intercalated P-120 and sodium intercalated P-120 has a positive thermocouple sensitivity that is close to the sum of the magnitudes of the absolute thermoelectric powers of the bromine intercalated P-120 and the sodium intercalated P-120. Fig. 21 shows the linear relationship of the measured voltage with the temperature difference between hot and cold points for the junction comprising bromine intercalated P-100 and sodium intercalated P-100.

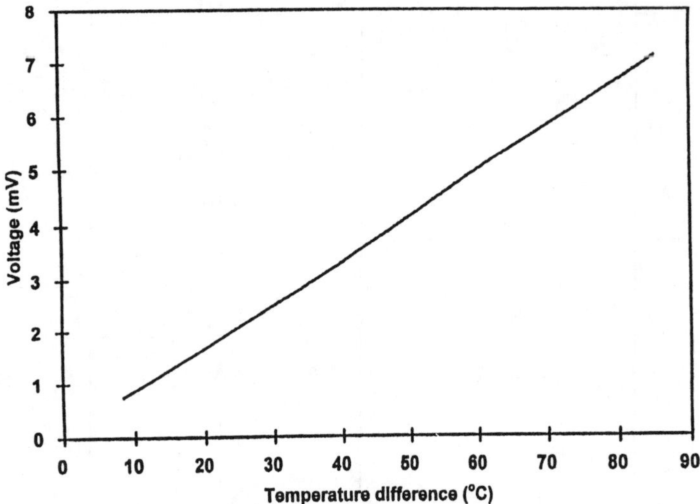

Fig. 21 *Variation of the measured voltage with the temperature difference between hot and cold points for the epoxy-matrix composite junction comprising bromine intercalated P-100 and sodium intercalated P-100 carbon fibers.*

A junction comprising n-type and p-type partners has a thermocouple sensitivity that is close to the sum of the magnitudes of the absolute thermoelectric powers of the two partners. This is because the electrons in the n-type partner as well as the holes in the p-type partner move away from the hot point toward the corresponding cold point. As a result, the overall effect on the voltage difference between the two cold ends is additive.

By using junctions comprising strongly n-type and strongly p-type partners, a thermocouple sensitivity as high as +82 μV/°C was attained. Semiconductors are known to exhibit much higher values of the Seebeck coefficient than metals, but the need to have thermocouples in the form of long wires makes metals the main materials for thermocouples. Intercalated carbon fibers exhibit much higher values of the Seebeck coefficient than metals. Yet, unlike semiconductors, their fiber form and fiber composite form make them convenient for practical use as thermocouples.

The thermocouple sensitivity of the carbon fiber epoxy-matrix composite junctions is independent of the extent of curing and is the same for unidirectional and crossply junctions. This is consistent with the fact that the thermocouple effect hinges on the difference in the bulk properties of the two partners, and is not an interfacial phenomenon. This behavior means that the interlaminar interfaces in a fibrous composite serve as thermocouple junctions in the same way, irrespective of the lay-up configuration of the dissimilar fibers in the laminate. As a structural composite typically has fibers in multiple directions, this behavior facilitates the use of a structural composite as a thermocouple array.

It is important to note that the thermocouple junctions do not require any bonding agent other than the epoxy, which serves as the matrix of the composite and does not serve as an electrical contact medium (since it is not conducting). In spite of the presence of the epoxy matrix in the junction area, direct contact occurs between a fraction of the fibers of a lamina and a fraction of the fibers of the other lamina, thus resulting in a conduction path in the direction perpendicular to the junction.

3.2.5 Polymer-matrix composites for reflecting or absorbing electromagnetic radiation

Polymer-matrix composites containing continuous carbon fibers are excellent for reflecting electromagnetic radiation such as radio waves [358-362], though carbon-matrix composites containing continuous carbon fibers are even better, due to the electrical conductivity of the carbon matrix [362]. The reflectivity renders the ability to shield EMI.

Absorption rather than reflection is attractive for structural composites used for aircraft that exhibits low obsrvability. The addition of carbonyl-iron particles that have been surface-treated with a silane coupler is effective due to the high permeability, high permittivity and low conductivity of the particles [363]. The low conductivity is due to the surface treatment.

4. Conclusion

Composite materials for electronic functions include non-structural materials (primarily for electronic packaging) and structural materials (primarily for smart structures). The former involves mainly polymer-matrix and metal-matrix composites; the latter involves mainly polymer-matrix and cement-matrix composites. The former involves mainly discontinuous reinforcements; the latter involves mainly continuous reinforcements in the case of polymer-matrix composites and mainly discontinuous reinforcements in the case of cement-matrix composites. The electronic functions make use of the electrical, electromagnetic and thermal properties of the composite materials.

Acknowledgement

This work was supported in part by Defense Advanced Research Projects Agency, U.S.A., and in part by National Science Foundation, U.S.A.

References

1. D.D.L. Chung, *Materials for Electronic Packaging,* Butterworth-Heinemann, Woburn, MA, 1995, pp. 3-4.
2. W. Eakin, K. Gardiner and J. Nayak, *J. Electronics Manuf.* 1, 13-22 (1991).
3. V. Sarihan and T. Fang, *Structural Analysis in Microelectronic and Fiber Optic Systems*, American Society of Mechanical Engineers, EEP, 1995, vol. 12, ASME, New York, NY, pp. 1-4.
4. W.E. Marsh, K. Kanakarajan and G.D. Osborn, *Polymer/Inorganic Interfaces*, Materials Research Soc. Symp. Proc., Materials Research Soc., Pittsburgh, PA, 1993, vol. 304, pp. 91-96.
5. L.T. Nguyen and I.C. Noyan, *Polymer Eng. Sci.,* 28(16), 1013-1025 (1988).
6. W.-J. Yang and K. Kudo, *Proc. Int. Symp. Heat Transfer Science and Technology*, Beijing, China, Hemisphere Publ. Corp., Washington, DC, 1988, pp. 14-30.
7. W.C. Riley and J. Kesapradist, Proc. *Technical Program: Nepcon West '92*, Cahners Exposition Group, Des Plaines, IL, 1992, vol. 1, pp.105-122.
8. A. Dasgupta, *6th Int. SAMPE Electronics Conf.,* 1992, pp. 782-794.
9. A. Zweben, *JOM* 50(6), 47-51 (1998).
10. H. Korner, A. Shiota and C.K. Ober, *ANTEC '96: Plastics–Racing into the Future, Conf. Proc.,* Society of Plastics Engineers, Technical Papers Series, no. 42, Brookfield, Conn., 1996, vol. 2, pp. 1458-1461.
11. G.S. Swei and D.J. Arthur, *3rd Int. SAMPE Symp. Exhib.,* 1989, SAMPE, Covina, CA, pp. 1111-1124.

12. L. Li and D.D.L. Chung, *Composites* 22(3), 211-218 (1991).

13. X. Lu and G. Xu, *J. Appl. Polymer Sci.* 65(13), 2733-2738 (1997).

14. D. Klosterman and L. Li, J. Electronics Manufacturing 5(4), 277-287 (1995).

15. S.K. Kang, R. Rai and S. Purushothaman, 1996 Proc. *46th Electronic Components & Technology,* IEEE, New York, NY, 1996, pp. 565-570.

16. G.F.C.M. Lijten, H.M. Van Noort and P.J.M. Beris, *J. Electronics Manufacturing* 5(4), 253-261 (1995).

17. M.P. Zussman, B. Kirayoglu, S. Sharkey and D.J. Powell, *6th Int. SAMPE Electronics Conf.*, 1992, pp. 437-448.

18. J.D. Bolt and R.H. French, *Adv. Mater. Processes* 134(1), 32-35 (1988).

19. J.D. Bolt, D.P. Button and B.A. Yost, *Mater. Sci. Eng.* A109, 207-211 (1989).

20. S.P. Mukerherjee, D. Suryanarayana and D.H. Strope, *J. Non-Crystalline Solids* 147 & 148, 783-791 (1992).

21. M.P. Zussman, B. Kirayoglu, S. Sharkey and D.J. Powell, *6th Int. SAMPE Electronics Conf.*, 1992, pp. 437-448.

22. J.J. Glatz, R. Morgan and D. Neiswinger, *Int. SAMPE Electronics Conf.,* 1992, vol. 6, pp. 131-145.

23. A. Bertram, K. Beasley and W. De La Torre, *Naval Engineers J.* 104(3), 276-285 (1992).

24. D. Brookstein and D. Maass, *7th Int. SAMPE Electronics Conf.,* 1994, pp. 310-327.

25. T.F. Fleming and W.C. Riley, *Proc. SPIE – The Int. Soc. for Optical Eng.,* Soc. Photo-Optical Instrumentation Engineers, Bellingham, WA, 1993, vol. 1997, pp. 136-147.

26. T.F. Fleming, C.D. Levan and W.C. Riley, *Proc. Technical Conf., Int. Electronics Packaging Conf.,* Wheaton, Ill, Int. Electronics Packaging Society, 1995, pp. 493-503.

27. A.M. Ibrahim, *6th Int. SAMPE Electronics Conf.,* 1992, pp. 556-567.

28. N. Kiuchi, K. Ozawa, T. Komami, O. Katoh, Y. Arai, T. Watanabe and S. Iwai, *Int. SAMPE Tech. Conf.,* vol. 30, 1998, SAMPE, Covina, CA, pp. 68-77.

29. J.W.M. Spicer, D.W. Wilson, R. Osiander, J. Thomas and B.O. Oni, *Proc. SPIE – the International Society for Optical Engineering*, vol. 3700, 1999, pp. 40-47.

30. D.A. Foster, *SAMPE Q.* 21(1), 58-64 (1989).

31. D.A. Foster, *34th Int. SAMPE Symp. Exhib.*, Book 2 (of 2), SAMPE, Covina, CA, 1989, pp. 1401-1410.

32. W. De La Torre, *6th Int. SAMPE Electronics Conf.*, 1992, pp. 720-733.

33. P.D. Wienhold, D.S. Mehoke, J.C. Roberts, G.R. Seylar and D.L. Kirkbride, *Int. SAMPE Tech. Conf.*, vol. 30, 1998, SAMPE, Covina, CA, pp. 243-255.

34. X. Luo and D.D.L. Chung, *Composites*: Part B 30(3), 227-231 (1999).

35. L.G. Morin Jr. and R.E. Duvall, *Proc. Int. SAMPE Symp. Exhib.*, vol. 43, No. 1, 1998, SAMPE, Covina, CA, pp. 874-881.

36. G. Lu, X. Li and H. Jiang, *Composites Sci. Tech.* 56, 193-200 (1996).

37. P.L. Smaldone, *27th Int. SAMPE Technical Conf.*, 1995, pp. 819-829.

38. J.J. Glatz, D.L. Vrable, T. Schmedake and C. Johnson, *6th Int. SAMPE Electronics Conf.*, 1992b, p. 334-346.

39. R. Crossman, *Northcon/85 – Conf. Rec.*, distributed by Western Periodicals Co., North Hollywood, CA, published by Electronic Conventions Management Inc., Los Angeles, CA, 1985, 21 pp.

40. L. Li and J.E. Morris, *Int. J. Microelectronic Packaging Mater. Tech.* 1(3), 159-175 (1998).

41. D.S. McLachlan, M. Blaszkiewicz and R.E. Newnham, *J. Am. Ceram. Soc.* 73(8), 2187-2203 (1990).

42. Z. Ziembik, M. Zabkowska-Waclawek and W. Waclawek, *J. Mater. Sci.* 34(14), 3495-3504 (1999).

43. M.G. Alexander, *Mater. Res. Bull.* 34(4), 603-611 (1999).

44. M. El Malhi, M.E. Achour, F. Lahjomri and Y. Bensalah, *J. Mater. Sci. Lett.* 18(8), 613-616 (1999).

45. G.J. Lee, K.D. Suh and S.S. Im, *Polymer Eng. Sci.* 38(3), 471-477 (1998).

46. S. Nakamura, T. Tomimura and G. Sawa, *Conf. Electrical Insulation and Dielectric Phenomena (CEIDP)*, Annual Report, vol. 1, 1998, IEEE, Piscataway, NJ, p. 265-268.

47. K.P. Sau, T.K. Chaki and D. Khastgir, *J. Appl. Polymer Sci.* 71(6), 887-895 (1999).

48. A.P. Burden, R.D. Forrest, S. Ravi and P. Silva, *J. Mater. Sci. Lett.* 17(17), 1467-1470 (1998).

49. C. Marchand, F. Pla and J.L. Greffe, *Proc. Int. Conf. Electric Charge in Solid Insulators*, CSC, 1998, p. 363-371.

50. V.M. Rudoi, O.V. Yaroslavtseva, T.N. Ostanina, L.P. Yurkina and O. Yu Subbotina, *Zashchita Metallov.* 34(5), 470 (1998).

51. L. Li and D.D.L. Chung, *Polymer Composites* 14(6), 467-472 (1993).

52. L. Li, P. Yih and D.D.L. Chung, *J. Electron. Mater.* 21(11), 1065-1071 (1992).

53. B. Guerrero, C. Alemán and R. Garza, *J. Polymer Eng.* 17(2), 95-110 (1997-1998).

54. M. Omastová, J. Pavlinec, J. Pionteck and F. Simon, *Polymer Int.* 43(2), 109-116 (1997).

55. X.B. Chen and J.-P. Issi, M. Cassart, J. Devaux and D. Billaud, *Polymer* 35(24), 5256-5258 (1994).

56. P. Chen and D.D.L. Chung, *J. Electron. Mater.* 24(1), 47-51 (1995).

57. J. Fournier, G. Boiteux, G. Seytre and G. Marichy, *J. Mater. Sci. Lett.* 16(20), 1677-1679 (1997).

58. A.R. Duggal and L.M. Levinson, *Electrically Based Microstructural Characterization II*, Materials Research Soc. Symp. Proc., vol. 500, 1998, MRS, Warrendale, PA, p. 247-252.

59. S. Hirano, A. Kishimoto and M. Miyayama, *J. Mater. Sci. Lett.* 17(13), 1133-1135 (1998).

60. A. Patnaik, Z. Zhu, G. Yang and Y. Sun, *Physica Status Solidi A – Applied Research* 169(1), 115-125 (1998).

61. L. Rector and H. Hyatt, *Annual Tech. Conf. – ANTEC*, Conf. Proc., vol. 2, 1998, Soc. Plast. Eng., Brookfield, CT, p. 1381-1383.

62. D.A. Olivero and D.W. Radford, *J. Reinf. Plastics Composites* 17(8), 674-690 (1998).

63. D. Klosterman and L. Li, *J. Electronics Manufacturing* 5(4), 277-287 (1995).

64. S.K. Kang, R. Rai and S. Purushothaman, 1996 Proc. *46th Electronic Components & Technology,* IEEE, New York, NY, 1996, pp. 565-570.

65. J. Miragliotta, R.C. Benson, T.E. Phillips and J.A. Emerson, *Electronic Packaging Materials Science X*, Materials Research Soc. Symp. Proc., vol. 515, 1998, MRS, Warrendale, PA, p. 245-250.

66. W.R. Lambert and W.H. Knausenberger, *Proceedings of the Technical Program – National Electronic Packaging and Production Conference* 3, 1512-1526 (1991).

67. J.A. Fulton, D.R. Horton, R.C. Moore, W.R. Lambert and J.J. Mottine, *Proc. 39th Electronic Components Conf.*, IEEE, 1989, p., 71-77.

68. W.R. Lambert, J.P. Mitchell, J.A. Suchin and J.A. Fulton, *Proc. 39th Electronic Components Conf.,* IEEE, 1989, p. 99-106.

69. P.B. Hogerton, J.B. Hall, J.M. Pujol and R.S. Reylek, *Mat. Res. Soc. Symp. Proc.* 154, 415 (1989).

70. L. Li and D.D.L. Chung, *J. Electronic Packaging* 119(4), 255 (1997).

71. P.T. Robinson, V. Florescu, G. Rosen and M.T. Singer, *Annual Connector & Interconnection Technology Symposium*, 1990, International Institute of Connector and Interconnection Technology, Deerfield, IL, p. 507-515.

72. G. Rosen, P.T. Robinson, V. Florescu and M.T. Singer, *Proc. 6th IEEE Holm Conf. Electrical Contacts and 15th Int. Conf. Electric Contacts*, IEEE, Piscataway, NJ, 1990, 151-165.

73. D.D. Johnson, *Proc. 3td Int. Symp. Advanced Packaging Materials: Processes, Properties and Interfaces*, IEEE, Piscataway, NJ and IMAPS, Reston, VA, 1997, p. 29-30.

74. T. Kokogawa, H. Morishita, K. Adachi, H. Otsuki, H. Takasago and T. Yamazaki, *Conference Record of 1991 International Display Confrence,* San Diego, 1991, IEEE, Piscataway, NJ and Society for Information Display, Playa Del Rey, CA, 1991, p. 45-48.

75. N.P. Kreutter, B.K. Grove, P.B. Hogerton and C.R. Jensen, *7th Electronic Materials and Processing Congress*, Cambridge, MA, Aug. 1992.

76. H. Yoshigahara, Y. Sagami, T. Yamazaki, A. Burkhart and M. Edwards, *Proc. Technical Program: National Electronic Packaging and Production Conf.,* NEPCON West '91, Cahners Exposition Group, Des Plaines, IL 1, 213-219 (1991).

77. D.M. Bruner, *Int. J. Microcircuite and Electronic Packaging* 18(3), 311 (1995).

78. J.J. Crea and P.B. Hogerton, *Proc. Technical Program: National Electronic Packaging and Production Conf., NEPCON West '91*, Cahners Exposition Group, Des Plaines, IL, 1991, Vol. 1, p. 251-259.

79. K. Gilleo, *Proc. Electricon '94*, Electronics Manufacturing Productivity Facility, Indianapolis, IN, 1994, pp. 11/1-11/12.

80. K. Chung, G. Dreier, P. Fitzgerald, A. Boyle, M. Lin and J. Sager, *Proc. 41st Electronic Components Conf.*, May 1991, p. 345.

81. J.C. Bolger and J.M. Czarnowski, *1995 Japan IEMT Symp. Proc. 1995 Int. Electronic Manufacturing Technology Symp.,* IEEE, New York, NY, 1996, p. 476-481.

82. Y. Xu and D.D.L. Chung, *J. Electron. Mater.* 28(11), 1307-1313 (1999).

83. G.Y. Chin, *Adv. Mater. Proc.* 137(1), 47, 50, 86 (1990).

84. S. Liang, S.R. Chong and E.P. Giannelis, *Proc. Electronic Components & Technology Conf.,* IEEE, New York, NY, vol. 48, 1998, pp. 171-175.

85. V. Agarwal, P. Chahal, R. R. Tummala and M.G. Allen, *Proc. Electronic Components and Technology Conf.,* IEEE, New York, NY, vol. 48, 1998, pp. 165-170.

86. T. Thongvigitmanee and G.S. May, Proc. SPIE – The Int. Soc. for Optical Engineering, vol. 3582, 1998, SPIE, Bellingham, WA, p. 79-83.

87. J.Y. Park and M.G. Allen, *Conf. Proc. – IEEE Applied Power Electronics Conf. and Exposition – APEC*, 1997, vol. 1, pp. 361-367.

88. J.Y. Park, L.K. Lagorce and M.G. Allen, *IEEE Trans. Magnetics* 33(5), pt. 1, 3322-3324 (1997).

89. J.Y. Park and M.G. Allen, *Int. J. Microcircuits & Electronic Packaging* 21(3), 243-252 (1998).

90. A. Gruskova, J. Slama, R. Vicen, L. Simekova, I. Hudec and A. Klanduchova, *J. Physique IV* 8(2), 421-424 (1998).

91. R. Charbonneau, *Proc. Int. SAMPE Symp. Exhib.,* vol. 43, No. 1, 1998, SAMPE, Covina, CA, pp. 833-844.

92. J. Wang, V.V. Varadan and V.K. Varadan, *SAMPE J.* 32(6), 18-22 (1996).

93. S. Maugdal and S. Sankaran, *Recent Trends in Carbon*, Proc. National Conf., ed. O.P. Bahl, Shipra Publ., Delhi, India, 1997, p. 12-19.

94. J.T. Hoback and J.T. Reilly, *J. Elastomers Plastics* 20(1), 54-69 (1988).

95. G. Lu, X. Li and H. Jiang, *Composites Sci. Tech.* 56, 193-200 (1996).

96. D.W. Radford, *J. Adv. Mater.*, Oct. 1994, pp. 45-53.

97. C. Huang and J. Pai, *J. Appl. Polymer Sci.* 63(1), 115-123 (1997).

98. D.W. Radford and B.C. Cheng, *J. Testing & Evaluation* 21(5), 396-401 (1993).

99. S.R. Gerteisen, *Northcon/85 – Conf. Rec.,* distributed by Western Periodicals Co., North Hollywood, CA, Pap. 6.1, published by Electronic Conventions Management Inc., Los Angeles, CA, 1985, 9 pp.

100. P. O'Shea, *Evaluation Engineering* 34(8), 84-93 (1995).

101. P. O'Shea, *Evaluation Engineering* 35(8), 56-61 (1996).

102. R.A. Rothenberg and D.C. Inman, *1994 Int. Symp. Electromagnetic Compatibility*, The Technical Group on EMC of the Institute of Electronics, Information and Communication Engineers, and The Technical Group on EMC of the Institute of Electrical Engineers of Japan, 1994, p. 818.

103. J.W.M. Child, *Electronic Production,* Oct. 1986, p. 41-47.

104. A.K. Subramanian, D.C. Pande and K. Boaz, *Proc. 1995 Int. Conf. Electromagnetic Interference and Compatibility*, Soc. EMC Engineers, Madras, India, 1995, p. 139-147.

105. W. Hoge, *Evaluation Engineering* 34(1), 84-86 (1995).

106. J.F. Walther, *IEEE 1989 Int. Symp. Electromagnetic Compatibility: Symp. Record*, IEEE, New York, NY, 1989, p. 40-45.

107. H.W. Denny and K.R. Shouse, *IEEE 1990 Int. Symp. Electromagnetic Compatibility: Symp. Record,* IEEE, New York, NY, 1990, p. 20-24.

108. R. Bates, S. Spence, J. Rowan and J. Hanrahan, *8th Int. Conf. Electromagnetic Compatibility,* Electronics Division, Insitution of Electrical Engineers, London, 1992, p. 246-250.

109. J.A. Catrysse, *8th Int. Conf. Electromagnetic Compatibility,* Electronics Division, Institution of Electrical Engineers, London, 1992, p. 251-255.

110. A.N. Faught, *IEEE Int. Symp. Electromagnetic Compatibility,* IEEE, New York, NY, 1982, p. 38-44.

111. G. Kunkel, *IEEE Int. Symp. Electromagnetic Compatibility,* IEEE, New York, NY, 1980, p. 211-216.

112. K.P. Sau, T.K. Chaki, A. Chakraborty and D. Khastgir, *Plastics Rubber & Composites Processing & Applications* 26(7), 291-297 (1997).

113. X. Shui and D.D.L. Chung, *J. Electron Mater*. 24(2), 107-113 (1995).

114. X. Shui and D.D.L. Chung, *J. Electron. Mater*. 25(6), 930-934 (1996).

115. X. Shui and D.D.L. Chung, *J. Electron. Mater*. 26(8), 928-934 (1997).

116. X. Shui and D.D.L. Chung, *J. Electron. Packaging* 119(4), 236-238 (1997).

117. X. Lu, G. Xu, P.G. Hofstra and R.C. Bajcar, *J. Polymer Sci., Part B – Polymer Physics* 36(13), 2259-2265 (1998).

118. Y. Xu and D.D.L. Chung, *Composite Interfaces*, in press.

119. Y. Xu and D.D.L. Chung, *J. Electron. Packaging*, in press.

120. P.K. Rohatgi, *Defence Science J.* 43(4), 323-349 (1993).

121. R. Asthana, *Adv. Performance Mater.* 5(3), 213-255 (1998).

122. M.A. Occhionero, R.A. Hay, R.W. Adams and K.P. Fennessy, *Proc. 1998 Int. Symp. Microelectronics*, 1998, pp. 687-692.

123. R.F. Jeffries and D.J. Borrows, *GEC J. Research* 11(3), 150-166 (1994).

124. J.L. Bugeau, K.M. Heitkamp and D. Kellerman, *IEEE MTT-S Int. Microwave Symp. Digest* 3, 1575-1578 (1995).

125. B.E. Novich, R.W. Adams and K.P. Fennessy, *Proc. 1995 Int. Electronics Packaging Conf.,* Int. Electronics Packaging Soc., Wheaton, IL, 1995, pp. 94-101.

126. G.L. Romero, J.M. Fusaro, J.L. Martinez, Jr., *Conf. Record – IAS Annual Meeting*, IEEE Industry Applications Society, 1995, pp. 916-922.

127. G.L. Romero and J.L. Martinez, Jr., *Int. J. Microcircuits & Electronic Packaging*, 1995, 18(3), 246-251.

128. P. Heller, EEP vol. 4-1, *Advances in Electronic Packaging,* ASME, 1993, pp. 429-433.

129. P. Heller, *Proc. Tech. Program – National Electronic Packaging and Production Conf.,* 1995, Vol. 1, Cahner Exposition Group, Des Plaines, IL, pp. 265-272.

130. J.L. Martinez, Jr., J.M. Fusaro, G.L. Romero, D. Humbert and D. Hooser, *Materials and Process Challenges: Aging Systems, Affordability, Alternative Applications,* SAMPE, Covina, CA 41(2), 1417-1426 (1996).

131. K. G. Beasley, *Int. SAMPE Electronics Conf.,* 1992, vol. 6, pp. 568-577.

132. R.D. Cottle, X. Chen, R.K. Jain, Z. Eliezer, L. Rabenberg and M.E. Fine, *JOM* 50(6), 67-69 (1998).

133. W.M. Peck, *American Society of Mechanical Engineers, Heat Transfer Division, (Publication) HTD*, vol. 329, No. 7, 1996, ASME, New York, NY, pp. 245-253.

134. A.L. Geiger and M. Jackson, *Adv. Mater. Proc.* 136(1), 6 pp (1989).

135. S. Lai and D.D.L. Chung, *J. Mater. Sci.* 29, 3128-3150 (1994).

136. X. Yu, G. Zhang and R. Wu, *Cailiao Gongcheng/J. Mater. Eng.*, June 1994, No. 6, 9-12.

137. B.E. Novich and R.W. Adams, Proc. *1995 Int. Electronics Packaging Conf.,* Int. Electronics Packaging Society, Wheaton, IL, 1995, pp. 220-227.

138. X.F. Yang and X.M. Xi, *J. Mater. Res.* 10(10), 2415-2417 (1995).

139. J.A. Hornor and G.E. Hannon, *6th Int. Electronics Packaging Conf.*, 1992, pp. 295-307.

140. Y.-L. Shen, A. Needleman and S. Suresh, *Met. Mater. Trans.* 25A(4), 839-850 (1994).

141. Y.-L. Shen, *Mater. Sci. Eng. A*, Sept. 1998, (2), 269-275.

142. J. Chiou and D.D.L. Chung, *J. Mater. Sci.* 28, 1435-1446 (1993).

143. J. Chiou and D.D.L. Chung, *J. Mater. Sci.* 28, 1447-1470 (1993).

144. J. Chiou and D.D.L. Chung, *J. Mater. Sci.* 28, 1471-1487 (1993).

145. M.E. Smagorinski, P.G. Tsantrizos, S. Grenier, A. Cavasin, T. Brzezinski and G. Kim, *Mater. Sci. Eng. A*, Mar. (1), 86-90 (1998).

146. K. Schmidt, C. Zweben and R. Arsenault, *ASTM Special Technical Publication*, Philadelphia, PA, 1990, No. 1080, pp. 155-164.

147. Y. Yasutomi, J. Sawada, T. Kikuchi, K. Nakamura, Y. Manabe, K. Nagano, H. Kuroda, T. Sumi, H. Kubokawa, M. Nagai, H. Kogure, Y. Sawai and T. Kishi, *J. Mater. Sci.* 34(7), 1583-1593 (1999).

148. H. Jeong, D.K. Hsu and P.K. Liaw, *Composites Sci. Tech.* 58(1), 65-76 (1998).

149. L.M. Tham, M. Gupta and L. Cheng, *J. Mater. Proc. Tech.* 89-90, 128-134 (1999).

150. R. Saraswati and F.J. Polese, *Proc. SPIE – The Int. Soc. for Optical Engineering,* vol. 3582, 1998, SPIE, Bellingham, WA, p. 681-686.

151. S. Lai and D.D.L. Chung, *J. Mater. Sci.* 29, 2998-3016 (1994).

152. S. Lai and D.D.L. Chung, *J. Mater. Chem.* 6, 469-477 (1996).

153. S. Lai and D.D.L. Chung, *J. Mater. Sci.* 29, 6181-6198 (1994).

154. S. Wagner, M. Hook and E. Perkoski, *Proc. Electricon '93*, Electronics Manufacturing Productivity Facility, 1993, p. 11/1-11/14.

155. O.J. Ilegbusi, *(Paper) ASME*, no. 93-WA/EEP-28, 1993, pp. 1-7.

156. M.A. Lambert and L.S. Fletcher, HTD vol. 292, *Heat Transfer in Electronic Systems*, ASME, New York, NY, 1994, pp. 115-122.

157. M.A. Lambert and L.S. Fletcher, J. *Heat Transfer – Trans. ASME* 118(2), 478-480 (1996).

158. M.A. Kinna, *6th Int. SAMPE Electronics Conf.* pp., 547-555 (1992).

159. J.-M. Ting, M.L. Lake and D.R. Duffy, *J. Mater. Res.* 10(6), 1478-1484 (1995).

160. T.F. Fleming, C.D. Levan and W.C. Riley, *Proc. Technical Conf., Int. Electronics Packaging Conf.,* Wheaton, Ill, Int. Electronics Packaging Society, 1995, pp. 493-503.

161. T.F. Fleming and W.C. Riley, *Proc. SPIE – The Int. Soc. for Optical Eng.,* Soc. Photo-Optical Instrumentation Engineers, Bellingham, WA, 1993, vol. 1997, pp. 136-147.

162. H. Wang and S.H.J. Lo, *J. Mater. Sci. Lett.* 15(5), 369-371 (1996).

163. R.E. Morgan, S.L. Ehlers and J. Sosniak, *6th Int. SAMPE Electronics Conf.* pp. 320-333 (1992).

164. S. Yoo, M.S. Krupashankara, T.S. Sudarshan and R.J. Dowding, *Mater. Sci. Tech.* 14(2), 170-174 (1998).

165. Anonymous, *Metal Powder Report* no. 4, 28-31 (1997).

166. T.W. Kirk, S.G. Caldwell and J.J. Oakes, *Particulate Materials and Processes, Advances in Powder Metallurgy,* Publ. By Metal Powder Industries Federation, Princeton, NJ, vol. 9, 1992, pp. 115-122.

167. P. Yih and D.D.L. Chung, *J. Electron. Mater.* 24(7), 841-851 (1995).

168. A. Ward and D.J. Nelson, *Thermomechanical Phenomena in Electronic Systems* – Proc. Intersociety Conf. 1998, IEEE, Piscataway, NJ, p. 520-523.

169. S. Jha, *1995 Proc. 45th Electronic Components & Technology Conf.,* IEEE, New York, NY, 1995, pp. 542-547.

170. R. Chanchani and P.M. Hall, *IEEE Trans. Components, Hybrids, and Manufacturing Technology* 13(4), 743-750 (1990).

171. C. Woolger, *Materials World* 4(6), 332-333 (1996).

172. R. Chanchani and P.M. Hall, *IEEE Trans. Components, Hybrids, and Manufacturing Technology* 13(4), 743-750 (1990).

173. J.R. Hanson and J.L. Hauser, *Electronic Packaging & Production* 26(11), 48-51 (1986).

174. J. Korab, G. Korb and P. Sebo, *Proc. IEEE/CPMT Int. Electronics Manufacturing Technology (IEMT) Symp.* IEEE, Piscataway, NJ, pp. 104-108 (1998).

175. J. Korab, G. Korb, P. Stefanik and H.P. Degischer, *Composites*, Part A 30(8), 1023-1026 (1999).

176. G. Korb, J. Koráb and G. Groboth, Composites, Part A 29A, 1563-1567 (1998).

177. M.A. Kinna, *6th Int. SAMPE Electronics Conf.* pp., 547-555 (1992).

178. M.A. Lambert and L.S. Fletcher, HTD vol. 292, *Heat Transfer in Electronic Systems*, ASME, New York, NY, 1994, pp. 115-122.

179. C.H. Stoessel, C. Pan, J.C. Withers, D. Wallace and R.O. Loutfy, *Mater. Res. Soc. Symp. Proc.*, vol. 390 (Electronic Packaging Materials Science VII), 1995, pp. 147-152.

180. Y. LePetitcorps, J.M. Poueylaud, L. Albingre, B. Berdeu, P. Lobstein and J.F. Silvain, *Key Engineering Materials* 127-131, 327-334 (1997).

181. K. Prakasan, S. Palaniappan and S. Seshan, *Composites,* Part A 28(12), 1019-1022 (1997).

182. P. Yih and D.D.L. Chung, *Int. J. Powder Met.* 31(4), 335-340 (1995).

183. M. Ruhle, *Key Engineering Materials* 116-117, 1-40 (1996).

184. M. Ruhle, *J. European Ceramic Soc.* 16(3), 353-365 (1996).

185. K. Prakasan, S. Palaniappan and S. Seshan, *Composites,* Part A 28(12), 1019-1022 (1997).

186. J.A. Kerns, N.J. Colella, D. Makowiecki and H.L. Davidson, *Int. J. Microcircuits and Electronic Packaging* 19(3), 206-211 (1996).

187. P. Yih and D.D.L. Chung, *J. Mater. Sci.* 32(7), 1703-1709 (1997).

188. P. Yih and D.D.L. Chung, *J. Mater. Sci.* 32(11), 2873-2894 (1997).

189. C.H. Stoessel, J.C. Withers, C. Pan, D. Wallace and R.O. Loutfy, *Surf. Coatings Tech.* 76-77**,** 640-644 (1995).

190. Y.Z. Wan, Y.L. Wang, G.J. Li, H.L. Luo and G.X. Cheng, *J. Mater. Sci. Lett.* 16, 1561-1563 (1997).

191. D.A. Foster, *34th Int. SAMPE Symp. Exhib.,* Book 2 (of 2), SAMPE, Covina, CA, 1989, pp. 1401-1410.

192. Y. Feng, M. Ying and C. Wang, *Kang T'Ieh/Iron & Steel (Peking)* 33(10), 38-42 (1998).

193. H.L. Davidson, N.J. Colella, J.A. Kerns and D. Makowiecki, *1995 Proc. 45th Electronic Components & Technology Conf.,* IEEE, New York, NY, 1995, pp. 538-541.

194. K. Prakasan, S. Palaniappan and S. Seshan, *Composites,* Part A 28(12), 1019-1022 (1997).

195. H. Mavoori and S. Jin, *JOM* 50(6), 70-72 (1998).

196. R.D. Stevenson, W.J. Whatley, J.J. Glatz and J.W. McCoy, *P/M in aerospace, defense, and demanding applications – 1993 Proc. 3rd Int. Conf. Powder Met. In Aerospace, Defense, and Demanding Applications*, The Federation Princeton, NJ, 1993, pp. 269-276.

197. T. Parsonage, *7th Int. SAMPE Electronics Conf.*, 1994, pp. 280-295.

198. T.B. Parsonage, *National Electronic Packaging and Production Conf. – Proc. Technical Program (West and East)*, 1996, Reed Exhibition Companies, Norwalk, CT, pp. 325-334.

199. A.W. Gibson, S. Choi, T.R. Bieler and K. N. Subramanian, *Proc. 1997 IEEE Int. Symp. Electronics and the Environment,* ISEE and IEEE, New York, NY, 1997, pp. 246-251.

200. Anonymous, *IBM Tech. Disclosure Bull.* 29(4), 1573-1574 (1986).

201. S. Trombert, J. Chazelas, P., Bonniau, W. Van Moorleghem, M. Chandrasekaran and J.F. Silvain, *3rd Int. Conf. Intelligent Materials: Third European Conf. Smart Structures and Materials*, SPIE, Proc., vol. 2779, 1996, p. 475-480.

202. Y. Wu, J.A. Sees, C. Pouraghabagher, L.A. Foster, J.L. Marshall, E.G. Jacobs and R.F. Pinizzotto, *J. Electron. Mater.* 22(7), 769-777 (1993).

203. A.W. Gibson, S. Choi, T.R. Bieler and K. N. Subramanian, *Proc. 1997 IEEE Int. Symp. Electronics and the Environment,* ISEE and IEEE, New York, NY, 1997, pp. 246-251.

204. C.Y. Wang, Y.C. Chen and C.C. Lee, *IEEE Trans. Components, Hybrids, and Manufacturing Tech.* 16(8), 789-793 (1993).

205. C.T. Ho and D.D.L. Chung, *J. Mater. Res.* 5(6), 1266-1270 (1990).

206. M. Zhu and D.D.L. Chung, *J. Mater. Sci.* 32, 5312-5333 (1997).

207. K.M. Kearns, *Proc. Int. SAMPE Symp. Exhib.*, SAMPE, Covina, CA, 43(2), 1362-1369 (1998).

208. W.H. Pfeifer, J.A. Tallon, W.T. Shih, B.L. Tarasen and G.B. Engle, *6th Int. SAMPE Electronics Conf.,* 1992, pp. 734-747.

209. E. Lee and W.E. Davis, *Int. SAMPE Symp. Exhib.* (Proc.), vol. 43, no. 1, 1998, SAMPE, Covina, CA, p. 511-516.

210. W.T. Shih, F.H. Ho and B.B. Burkett, *7th Int. SAMPE Electronics Conf.*, 1994, pp. 296-309.

211. W. Kowbel, X. Xia and J.C. Withers, 43rd Int. SAMPE Symp., 1998, pp. 517-527.

212. W. Kowbel, X. Xia, C. Bruce and J.C. Withers, *Mater. Res. Soc. Symp. Proc.,* vol. 515 (Electronic Packaging Materials Science X), Warrendale, PA, 1998, pp. 141-146.

213. S.K. Datta, S.M. Tewari, J.E. Gatica, W. Shih and L. Bentsen, *Met. Mater. Trans. A,* 1999, 30A, 175-181.

214. W.T. Shih, *Proc. Int. Electronics Packaging Conf.,* 1995, IEPS, Wheaton, IL, pp. 211-219.

215. J.-M. Ting and M.L. Lake, *Diamond and Related Materials* 3(10), 1243-1248 (1994).

216. M.J. Montesano, *Mat. Tech.* 11(3), 87-91 (1996).

217. J. Klett, C. Walls and T. Burchell, *Ext. Abstr. Program–24th Bienn. Conf. Carbon,* 1999, pp. 132-133.

218. L. Hozer, Y.-M. Chiang, S. Ivanova and I. Bar-On, *J. Mater. Res.* 12(7), 1785-1789 (1997).

219. P.N. Kumta, *J. Mater. Sci.* 31(23), 6229-6240 (1996).

220. P.N. Kumta, T. Mah, P.D. Jero and R.J. Kerans, Mater. Lett. 21(3-4), 329-333 (1994).

221. J.Y. Kim and P.N. Kumta, Proc. *IEEE 1998 National Aerospace and Electronics Conf.,* NAECON 1998, celebrating 50 years, IEEE, New York, NY, 1998, pp. 656-665.

222. G.V. Chandrashekhar and M.W. Shafer, *J. Mater. Sci.* 24(9), 3353-3355 (1989).

223. P. Chen and D.D.L. Chung, *J. Electron. Mater.* 24(1), 47-51 (1995).

224. X. Fu and D.D.L. Chung, *Carbon* 36(4), 459-462 (1998).

225. D.W. Kellerman, *Mater. Res. Soc. Symp. Proc.,* vol. 372 (Hollow and Solid Spheres and Microspheres: Science and Technology Associated with Their Fabrication and Application), Warrendale, PA, 1995, pp. 239-245.

226. K. Biswas and G. S. Upadhyaya, *Mater. Design* 19(5-6), 231-240 (1998).

227. J. W. Newman, *Int. SAMPE Symp. Exhib.*, vol. 32 (1987), SAMPE, Covina, CA, pp. 938-944.

228. S. Furukawa, Y. Tsuji and S. Otani, Proc. *30th Japan Congress on Materials Research* (1987), Soc. of Materials Science, Kyoto, Jpn., pp. 149-152.

229. K. Saito, N. Kawamura and Y. Kogo, *Advanced Materials: The Big Payoff*, National SAMPE Technical Conf., vol. 21 (1989), Covina, CA, pp. 796-802.

230. S. Wen and D.D.L. Chung, *Cen. Concr. Res.* 29(3), 445-449 (1999).

231. T. Sugama, L. E. Kukacka, N. Carciello and D. Stathopoulos, *Cem. Concr. Res.* 19(3), 355-365 (1989).

232. X. Fu, W. Lu and D.D.L. Chung, *Cem. Concr. Res.* 26(7), 1007-1012 (1996).

233. X. Fu, W. Lu and D.D.L. Chung, *Carbon* 36(9), 1337-1345 (1998).

234. Y. Xu and D.D.L. Chung, *Cem. Concr. Res.* 29(5), 773-776 (1999).

235. T. Yamada, K. Yamada, R. Hayashi and T. Herai, *Int. SAMPE Symp. Exhib.,* vol. 36 (1991), SAMPE, Covina, CA, pt. 1, pp. 362-371.

236. T. Sugama, L. E. Kukacka, N. Carciello and B. Galen, *Cem. Concr. Res.* 18(2), 290-300 (1988).

237. B. K. Larson, L. T. Drzal and P. Sorousian, *Composites* 21(3), 205-215 (1990).

238. A. Katz, V.C. Li and A. Kazmer, *J. Materials Civil Eng.* 7(2), 125-128 (1995).

239. S. B. Park and B. I. Lee, Cem. Concr. *Composites* 15(3), 153-163 (1993).

240. P. Chen, X. Fu and D.D.L. Chung, *ACI Mater. J.* 94(2), 147-155 (1997).

241. P. Chen and D.D.L. Chung, *Composites* 24(1), 33-52 (1993).

242. A. M. Brandt and L. Kucharska, *Materials for the New Millennium,* Proc. Mater. Eng. Conf., vol. 1 (1996), ASCE, New York, NY, pp. 271-280.

243. H. A. Toutanji, T. El-Korchi, R. N. Katz and G. L. Leatherman, *Cem. Concr. Res.* 23(3), 618-626 (1993).

244. N. Banthia and J. Sheng, *Cem. Concr. Composites* 18(4), 251-269 (1996).

245. H. A. Toutanji, T. El-Korchi and R. N. Katz, *Com. Con. Composites* 16(1), 15-21 (1994).

246. S. Akihama, T. Suenaga and T. Banno, *Int. J. Cem. Composites & Lightweight Concrete* 6(3), 159-168 (1984).

247. M. Kamakura, K. Shirakawa, K. Nakagawa, K. Ohta and S. Kashihara, *Sumitomo Metals* (1983).

248. A. Katz and A. Bentur, *Cem. Concr. Res.* 24(2), 214-220 (1994).

249. Y. Ohama and M. Amano, *Proc. 27th Japan Congress on Materials Research* (1983), Soc. Mater. Sci., Kyoto, Japan, pp. 187-191.

250. Y. Ohama, M. Amano and M. Endo, *Concrete Int.: Design & Construction* 7(3), 58-62 (1985).

251. K. Zayat and Z. Bayasi, *ACI Mater. J.* 93(2), 178-181 (1996).

252. P. Soroushian, F. Aouadi and M. Nagi, *ACI Mater. J.* 88(1), 11-18 (1991).

253. B. Mobasher and C. Y. Li, *ACI Mater. J.* 93(3), 284-292 (1996).

254. N. Banthia, A. Moncef, K. Chokri and J. Sheng, *Can. J. Civil Eng.* 21(6), 999-1011 (1994).

255. B. Mobasher and C. Y. Li, *Infrastructure: New Materials and Methods of Repair*, Proc. Mater. Eng. Conf., n 804 (1994), ASCE, New York, NY, pp. 551-558.

256. P. Soroushian, M. Nagi and J. Hsu, *ACI Mater. J.* 89(3), 267-276 (1992).

257. P. Soroushian, *Construction Specifier* 43(12), 102-108 (1990).

258. A. K. Lal, *Batiment Int./Building Research & Practice* 18(3), 153-161 (1990).

259. S. B. Park, B. I. Lee and Y. S. Lim, *Cem. Concr. Res.* 21(4), 589-600 (1991).

260. S. B. Park and B. I. Lee, *High Temperatures – High Pressures* 22(6), 663-670 (1990).

261. P. Soroushian, Mohamad Nagi and A. Okwuegbu, *ACI Mater. J.* 89(5), 491-494 (1992).

262. M. Pigeon, M. Azzabi and R. Pleau, *Cem. Concr. Res.* 26(8), 1163-1170 (1996).

263. N. Banthia, K. Chokri, Y. Ohama and S. Mindess, *Adv. Cem. Based Mater.* 1(3), 131-141 (1994).

264. N. Banthia, C. Yan and K. Sakai, *Com. Concr. Composites* 20(5), 393-404 (1998).

265. T. Urano, K. Murakami, Y. Mitsui and H. Sakai, *Composites – Part A: Applied Science & Manufacturing* 27(3), 183-187 (1996).

266. A. Ali and R. Ambalavanan, *Indian Concrete J.* 72(12), 669-675.

267. P. Chen, X. Fu and D.D.L. Chung, *Cem. Concr. Res.* 25(3), 491-496 (1995).

268. M. Zhu and D.D.L. Chung, *Cem. Concr. Res.* 27(12), 1829-1839 (1997).

269. M. Zhu, R. C. Wetherhold and D.D.L. Chung, *Cem. Concr. Res.* 27(3), 437-451 (1997).

270. P. Chen and D.D.L. Chung, *Smart Mater. Struct.* 2, 22-30 (1993)

271. P. Chen and D.D.L. Chung, *Composites*, Part B, 27B, 11-23 (1996).

272. P. Chen and D.D.L. Chung, *J. Am. Ceram. Soc.* 78(3), 816-818 (1995).

273. D.D.L. Chung, *Smart Mater. Struct.* 4, 59-61 (1995).

274. P. Chen and D.D.L. Chung, *ACI Mater. J.* 93(4), 341-350 (1996).

275. X. Fu and D.D.L. Chung, *Cem. Concr. Res.* 26(1), 15-20 (1996).

276. X. Fu, E. Ma, D.D.L. Chung and W. A. Anderson, *Cem. Concr. Res.* 27(6), 845-852 (1997).

277. X. Fu and D.D.L. Chung, *Cem. Concr. Res.* 27(9), 1313-1318 (1997).

278. X. Fu, W. Lu and D.D.L. Chung, *Cem. Concr. Res.* 28(2), 183-187 (1998).

279. S. Wen and D.D.L. Chung, *Cem. Concr. Res.*, in press.

280. Z. Shi and D.D.L. Chung, *Cem. Concr. Res.* 29(3), 435-439 (1999).

281. Q. Mao, B. Zhao, D. Sheng and Z. Li, *J. Wuhan U. Tech.*, Mater. Sci. Ed., 11(3), 41-45 (1996).

282. Q. Mao, B. Zhao, D. Shen and Z. Li, *Fuhe Cailiao Xuebao/Acta Materiae Compositae Sinica* 13(4), 8-11 (1996).

283. M. Sun, Q. Mao and Z. Li, *J. Wuhan U. Tech.*, Mater. Sci. Ed., 13(4), 58-61 (1998).

284. B. Zhao, Z. Li and D. Wu, *J. Wuhan Univ. Tech.,* Mater. Sci. Ed., 10(4), 52-56 (1995).

285. S. Wen and D.D.L. Chung, *Cem. Concr. Res.* 29(6), 961-965 (1999).

286. M. Sun, Z. Li, Q. Mao and D. Shen, *Cem. Concr. Res.* 28(4), 549-554 (1998).

287. M. Sun, Z. Li, Q. Mao and D. Shen, *Cem. Concr. Res.* 28(12), 1707-1712 (1998).

288. S. Wen and D.D.L. Chung, *Cem. Concr. Res.,* in press.

289. D. Bontea, D.D.L. Chung and G.C. Lee, *Cem. Concr. Res.,* in press.

290. S. Wen and D.D.L. Chung, *Cem. Concr. Res.,* in press.

291. J. Lee and G. Batson, *Materials for the New Millennium*, Proc. 4th Mater. Eng. Conf., vol. 2 (1996), ASCE, New York, NY, pp. 887-896.

292. X. Fu and D.D.L. Chung, *ACI Mater. J.* 96(4), 455-461 (1999).

293. Y. Xu and D.D.L. Chung, *Cem. Concr. Res.* 29(7), 1117-1121 (1999).

294. Y. Shinozaki, *Adv. Mater.: Looking Ahead to the 21st Century*, Proc. 22nd National SAMPE Tech. Conf. Vol. 22 (1990), SAMPE, Covina, CA, pp. 986-997.

295. X. Fu and D.D.L. Chung, *Cem. Concr. Res.* 25(4), 689-694 (1995).

296. J. Hou and D.D.L. Chung, *Cem. Concr. Res.* 27(5), 649-656 (1997).

297. G. G. Clemena, *Materials Performance* 27(3), 19-25 (1988).

298. R. J. Brousseau and G. B. Pye, *ACI Mater. J.* 94(4), 306-310 (1997).

299. P. Chen and D.D.L. Chung, *Smart Mater. Struct.* 2, 181-188 (1993).

300. X. Wang, Y. Wang and Z. Jin, *Fuhe Cailiao Xuebao/Acta Materiae Compositae Sinica* 15(3), 75-80 (1998).

301. N. Banthia, S. Djeridane and M. Pigeon, *Cem. Concr. Res.* 22(5), 804-814 (1992).

302. P. Xie, P. Gu and J. J. Beaudoin, *J. Mater. Sci.* 31(15), 4093-4097 (1996).

303. Z. Shui, J. Li, F. Huang and D. Yang, *J. Wuhan Univ. Tech.*, Mater. Sci. Ed., 10(4), 37-41 (1995).

304. X. Fu and D.D.L. Chung, *Cem. Concr. Res.* 28(6), 795-801 (1998).

305. X. Fu and D.D.L. Chung, *Cem. Concr. Res.* 26(10), 1467-1472 (1996); 27(2), 314 (1997).

306. T. Fujiwara and H. Ujie, *Tohoku Kogyo Daigaku Kiyo, 1: Rikogakuhen.* n 7, 179-188 (1987).

307. Y. Shimizu, A. Nishikata, N. Maruyama and A. Sugiyama, *Terebijon Gakkaishi/J. Inst. Television Engineers of Japan* 40(8), 780-785 (1986).

308. P. Chen and D.D.L. Chung, *ACI Mater. J.* 93(2), 129-133 (1996).

309. T. Uomoto and F. Katsuki, *Doboku Gakkai Rombun-Hokokushu/Proc. Japan Soc. Civil Engineers,* n 490, pt. 5-23, 167-174 (1994-1995).

310. C. M. Huang, D. Zhu, C. X. Dong, W. M. Kriven, R. Loh and J. Huang, *Ceramic Eng. Sci. Proc.* 17(4), 258-265 (1996).

311. C. M. Huang, D. Zhu, X. Cong, W. M. Kriven, R. R. Loh and J. Huang, *J. Am. Ceramic Soc.* 80(9), 2326-2332 (1997).

312. T-J. Kim and C-K. Park, *Cem. Concr. Res.* 28(7), 955-960 (1998).

313. S. Igarashi and M. Kawamura, *Doboku Gakkai Rombun-Hokokushu/Proc. Japan Soc. Civil Eng.*, n 502, pt 5-25, 83-92 (1994).

314. M. Z. Bayasi and J. Zeng, *ACI Structural J.* 94(4), 442-446 (1997).

315. G. Campione, S. Mindess and G. Zingone, *ACI Mater. J.* 96(1), 27-34 (1999).

316. T. Yamada, K. Yamada and K. Kubomura, *J. Composite Mater.* 29(2), 179-194 (1995).

317. S. Delvasto, A. E. Naaman and J. L. Throne, *Int. J. Cement Composites & Lightweight Concrete* 8(3), 181-190 (1986).

318. C. Park, Nippon Seramikkusu Kyokai *Gakujutsu Ronbunshi – J. Ceramic Soc. Japan* 106(1231), 268-271 (1998).

319. Y. Shao, S. Marikunte and S. P. Shah, *Concrete Int.* 17(4), 48-52 (1995).

320. S. Wen and D.D.L. Chung, *Cem. Concr. Res.,* in press.

321. S. Wen and D.D.L. Chung, *Cem. Concr. Res.*, in press.

322. W.F.A. Davies, *J. Phys.*, D: App. Phys. 7, 120-130 (1974).

323. W.J. Gadja, Report RADC-TR-78-158, A059029, 1978.

324. P. Li, W. Strieder and T. Joy, *J. Comp. Mater.* 16, 53-64 (1982).

325. T. Choi, P. Ajmera and W. Strieder, *J. Comp. Mater.* 14, 130-141 (1980).

326. T. Joy and W. Strieder, *J. Comp. Mater.* 13, 72-78 (1979).

327. K.W. Tse and C.A. Moyer, *Mater. Sci. and Eng.* 49, 41-46 (1981).

328. V. Volpe, *J. Comp. Mater.* 14, 189-198 (1980).

329. V.G. Shevchenko, A.T. Ponomarenko and N.S. Enikolopyan, *Int. J. Appl. Electromagnetics in Materials* 5(4), 267-277 (1994).

330. X.B. Chen and D. Billaud, *Ext. Abstr. Program – 20th Bienn. Conf. Carbon*, 1991, pp. 274-275.

331. X. Wang and D.D.L. Chung, *Composite Interfaces* 5(3), 191-199 (1998).

332. X. Wang and D.D.L. Chung, *Polymer Composites* 18(6), 692-700 (1997).

333. X. Wang and D.D.L. Chung, *Carbon* 35(5), 706-709 (1997).

334. X. Wang and D.D.L. Chung, *J. Mater. Res.* 14(11), 4224-4229 (1999).

335. S. Wang and D.D.L. Chung, *Composite Interfaces* 6(6), 497-506 (1999).

336. S. Wang and D.D.L. Chung *Polymer Composites*, in press.

337. X. Luo and D.D.L. Chung, *Composites Sci. Tech.*, in press.

338. N. Muto, H. Yanagida, T. Nakatsuji, M. Sugita, Y. Ohtsuka and Y. Arai, *Smart Mater. Struct.* 1, 324-329 (1992).

339. X. Wang, X. Fu and D.D.L. Chung, *J. Mater. Res.* 14(3), 790-802 (1999).

340. X. Wang and D.D.L. Chung, *Composites: Part B,* 29B(1), 63-73 (1998).

341. P.E. Irving and C. Thiogarajan, *Smart Mater. Struct.* 7, 456-466 (1998).

342. X. Wang, S. Wang and D.D.L. Chung, *J. Mater. Sci.* 34(11), 2703-2714 (1999).

343. S. Wang and D.D.L. Chung, *Polymer Composites*, in press.

344. N. Muto, H. Yanagida, M. Miyayama, T. Nakatsuji, M. Sugita and Y. Ohtsuka, J. *Ceramic Soc. Japan* 100(4), 585-588 (1992).

345. N. Muto, H. Yanagida, T. Nakatsuji, M. Sugita, Y. Ohtsuka, Y. Arai and C. Saito, *Adv. Composite Mater.* 4(4), 297-308 (1995).

346. R. Prabhakaran, *Experimental Techniques* 14(1), 16-20 (1990).

347. M. Sugita, H. Yanagida and N. Muto, *Smart Mater. Struct.* 4(1A), A52-A57 (1995).

348. A.S. Kaddour, F.A.R. Al-Salehi, S.T.S. Al-Hassani and M.J. Hinton, *Composites Sci. Tech.* 51, 377-385 (1994).

349. O. Ceysson, M. Salvia and L. Vincent, *Scripta Materialia* 34(8), 1273-1280 (1996).

350. K. Schulte and Ch. Baron, *Composites Sci. Tech.* 36, 63-76 (1989).

351. K. Schulte, *J. Physique IV*, Colloque C7, 3, 1629-1636 (1993).

352. J.C. Abry, S. Bochard, A. Chateauminois, M. Salvia and G. Giraud, *Composites Sci. Tech.* 59(6), 925-935 (1999).

353. A. Tedoroki, H. Kobayashi and K. Matuura, *JSME Int. J. Series A – Solid Mechanics Strength of Materials* 38(4), 524-530 (1995).

354. S. Hayes, D. Brooks, T. Liu, S. Vickers and G.F. Fernando, *Proc. SPIE – the Int. Soc. for Optical Engineering*, V. 2718 (Smart Structures and Materials 1996: Smart Sensing, Processing, and Instrumentation), SPIE, Bellingham, WA, 1996, pp. 376-384.

355. S. Wang and D.D.L. Chung, *Composites*: Part B, in press.

356. S. Wang and D.D.L. Chung, *Composites*: Part B, 30(6), 591-601 (1999).

357. S. Wang and D.D.L. Chung, *Composite Interfaces* 6(6), 519-530 (1999).

358. P.B. Jana and A.K. Mallick, *Elastomers and Plastics* 26(1), 58-73 (1994).

359. D.J. Atkins, J.A. Miller and J.D. Sanders, 8th Int. Conf. Electromagnetic Campatibility, Electronics Division, Institution of Electrical Engineering, London, 1992, pp. 100-107.

360. D.A. Jaworske, J.R. Gaier, C.C. Hung and B.A. Banks, *SAMPE Q.* 18(1), 9-14 (1986).

361. M-S. Lin and C.H. Chen, *IEEE Transactions on Electromagnetic Compatibility* 35(1), 21-27 (1993).

362. X. Luo and D.D.L. Chung, *Composites*: Part B, 30(3), 227-231 (1999).

363. M. Matsumoto and Y. Miyata, *NTT R&D* 48(3), 343-348 (1999).

continued overleaf

Advanced Reading

Experimental Techniques and Design in Composite Materials 3

Ed. Pierluigi Priolo

Proceedings of the Third Seminar on Experimental Techniques and Design in Composite Materials, Cagliari, Italy, October 1996

Key Engineering Materials Vol. 144

ISBN 0-87849-778-1
1998, 316 pp, CHF 130.00 / US$ 98.00 / GBP 54.00

Several things have changed in the world of composites in recent years. While many industries have reverted to metals for motor car bodies, composites have found new openings in other areas such as civil engineering, particularly because they offer a simple and rapid means of restoring damaged structures. Nevertheless, the main obstacle to their widespread use in a variety of promising fields, such as machine construction, continues to be the difficulty in treating concepts like heterogeneity and anisotropy. Many problems remain open and some of these are listed below:

- lack of literature data on basic mechanical properties
- response to particular loading conditions
- influence of the environment on short and long term behaviour
- simple models for the complex behaviour of laminates following initial failure
- numerical simulation of geometrically complex composites
- new techniques for describing the damage state.

The present book addresses subjects related mainly to the above topics, providing a substantial contribution to gaining a better understanding of composite materials behaviour.

CONTENTS SECTIONS:

Detailed information on this title – including the full table of contents – is available on the internet at http://www.ttp.net/titles/778.htm.

Trans Tech Publications Ltd

Brandrain 6
CH-8707 Uetikon-Zuerich
Switzerland

Fax: +41 (1) 922 10 33
e-mail: ttp@ttp.net
Web: http://www.ttp.net

High Temperature Ceramic Matrix Composites III

Eds. K. Niihara, K. Nakano, T. Sekino, E. Yasuda

Proceedings of the 3rd International Conference on High Temperature Ceramic Matrix Composites (HT-CMC 3), September 1998, Osaka, Japan

Key Engineering Materials Vols. 164-165
CSJ Series – Publications of the Ceramic Cociety of Japan – Volume 3

ISBN 0-87849-818-4
Publication Date: September 1998
484 pages, CHF 230.00 / US$ 165.00 / £ 96.00

High performance materials are needed in many thermomechanical applications, such as advanced jet engines or gas turbines, thermal protection of space planes, thermal engineering and nuclear fusion. Ceramic-matrix composites, reinforced with long fibers, are expected to be the most promising candidates for such applications. However, several key problems will have to be solved in the near future. These problems include, above all, the development of a superior thermomechanical performance and oxidation resistance for ceramic fibers and composites, highly efficient protection against corrosion, better fiber/matrix interface design and performance, and cost effective processing.

The third international conference series on High Temperature Ceramic Matrix Composites (HT-CMC 3) focused on composites containing carbon, oxides, non-oxides and intermetallic compounds. It also brought together research on complementary topics, such as: ceramic fibers, composite processing, design, fiber/matrix interface, mechanical and thermomechanical behaviors, environmental effects, modeling and applications.

This book therefore presents the most significant advances that have been made in this field during the past few years and will thus be an essential work for anyone who is concerned with Ceramic Matrix Composites.

CONTENTS

Detailed information on this title – including the complete table of contents – is available on the internet at http://www.ttp.net or through TTP's E-Mail Preview Service. For further information, please send an e-mail to 'preview@ttp.net' with the word 'help' as the body of the message.

Brandrain 6
CH-8707 Uetikon-Zuerich
Switzerland

Fax: +41 (1) 922 10 33
e-mail: ttp@ttp.net
Web: http://www.ttp.net